OXFORD MEDICAL PUBLICATIONS

Cancer Risk after Medical Treatment

Cancer Risk after Medical Treatment

EDITED BY

MICHEL P. COLEMAN

Epidemiologist
International Agency for Research on Cancer

OXFORD   NEW YORK   TOKYO
OXFORD UNIVERSITY PRESS
1991

Oxford University Press, Walton Street, Oxford OX2 6DP
Oxford New York Toronto
Delhi Bombay Calcutta Madras Karachi
Petaling Jaya Singapore Hong Kong Tokyo
Nairobi Dar es Salaam Cape Town
Melbourne Auckland
and associated companies in
Berlin Ibadan
Oxford is a trade mark of Oxford University Press

Published in the United States
by Oxford University Press, New York

© *the contributors listed on p. xv, 1991*

All rights reserved. No part of this publication may be reproduced, stored in a retrieval system, or transmitted, in any form or by any means, electronic, mechanical, photocopying, recording, or otherwise, without the prior permission of Oxford University Press

This book is sold subject to the condition that it shall not, by way of trade or otherwise, be lent, re-sold, hired out, or otherwise circulated without the publisher's prior consent in any form of binding or cover other than that in which it is published and without a similar condition including this condition being imposed on the subsequent purchaser

A catalogue record for this book is available from the British Library

Library of Congress Cataloging in Publication Data
Cancer risk after medical treatment / edited by Michel P. Coleman.
(Oxford medical publications)
Includes index.
1. Cancer–Etiology. 2. Cancer–Risk factors. 3. Iatrogenic diseases. I. Coleman, Michel P. II. Series.
[DNLM: 1. Drug Therapy–adverse effects. 2. Iatrogenic Disease–prevention & control. 3. Neoplasms–etiology. 4. Neoplasms–prevention & control. 5. Radiotherapy–adverse effects. 6. Risk Factors. QZ 202 C21545]   RC268.48.C37 1991   616.99'4071–dc20   91–3392
ISBN 0–19–261781–8

Set by Footnote Graphics, Warminster, Wiltshire
Printed and bound in Great Britain by
Courier International Ltd
East Kilbride, Scotland

For Liz and Matti

Preface

Adverse consequences of medical treatment are increasingly a subject of concern, both to the general public and the medical profession. There have been several recent episodes of serious and unexpected adverse effects, including carcinogenesis, following medical treatment. Examples include the teratogenic effect of thalidomide, the carcinogenic effect of unopposed oestrogen replacement therapy on the endometrium after menopause, and the occurrence of vaginal adenocarcinoma in young daughters of women given diethylstilboestrol in early pregnancy.

The patient rightly expects the risks associated with medical treatment to be reduced to a minimum, and to be fully informed about these risks before consenting to undergo major therapeutic regimes or surgical procedures. There is also increasing professional interest in the subject. A new journal, *Iatrogenics*, devoted to the study of 'physician-induced complications of care', will appear from 1991; the first world congress on 'Safety in Medical Practice' was held in 1990, and there is a new International Society for the Prevention of Iatrogenic Complications. The most recent reviews of cancer caused by medical treatment appeared in 1977 (Schmähl et al. 1977) and 1982 (Penn 1982). This book attempts to provide an up-to-date review of cancers arising after medical treatment, in the widest sense of the term, and the extent to which those cancers can be said to be caused—or prevented—by that treatment.

Many short-term unwanted effects of medical treatment are well known. Most occur within hours or weeks; they occur quite often, and in some cases, such as alopecia after chemotherapy, are sufficiently predictable for the patient to be warned, or for preventive measures to be adopted; and they are usually reversible. Even so, the detection and acceptance of gastric haemorrhage resulting from taking aspirin took more than 50 years, and systematic surveillance of adverse reactions to drugs only became widespread in the 1960s, after the outbreak of the congenital deformities caused by thalidomide.

Serious long-term side-effects of treatment such as cancer are less obviously linked to the treatment itself, both for the patient and the physician, and these side-effects are thus harder to identify and avoid. The difficulties involved in assessing the risk of malignancy after medical treatment are considerable, and the disease being treated may itself be associated with an increased cancer risk, as with ulcerative colitis and colon cancer. The latent period of cancer induction is at least several years, and may be 20 years or more: in these periods of time patients may move home, lose contact with the treating physician, or die from unrelated causes. The timescale of cancer production may be shorter if the treatment acts as a promoter or late-stage carcinogen, but cancers are infrequent, and those caused by treatment are rarely distinguishable in any given patient from cancers arising spontaneously in the general population. For these reasons, systematic ascertainment of all cancers in a group of treated patients is required to assess whether they develop more cancers than would be expected. This implies sustained and complete follow-up of large numbers of patients over many years to detect new cancers or deaths. The administrative systems required for this purpose are often complex, and much labour and expense is required to maintain the quality of the information they contain.

Despite these problems, cancer risk after various medical interventions has been assessed in a number of recent studies, and this book represents an attempt to provide clinicians with an overview of this body of work, and to place the observed cancer risks in their proper clinical context.

Many carcinogenic exposures such as radiation and alkylating agents are used mainly for diseases which are themselves serious or potentially fatal, and the risk of subsequent malignancy may be considered a price worth paying. In other cases, however, such as indolent tumours or disabling but rarely fatal diseases such as rheumatoid arthritis, the balance of risk may be harder to assess. Such an assessment requires an understanding of the size of cancer risk likely to be incurred by a particular form of treatment, and the extent to which it may vary with the age or sex of the patient, or with the particular cancer concerned. For example, radiotherapy for cancer of the cervix causes a small, early increase in leukaemia risk and a larger, delayed risk of rectal cancer, but a lifelong reduction in breast cancer risk, as well as the large reduction in mortality from cervical cancer for which it is given.

Not all treatment-related cancer risks are so complex, but such considerations now affect the choice of first-line therapy for many malignant tumours. One oncologist has remarked that whereas the measure of success in treating neoplastic disease used to be survival of the patient, it is increasingly coming to be survival without development of a new malignancy. Such a statement could only be made in the context of major improvements in survival from tumours such as Hodgkin's disease, and acute lymphatic leukaemia in childhood, where only recently have significant proportions of patients survived beyond the likely induction periods of chemical and radiation carcinogenesis; but it serves as a reminder that assessment of the balance between risks and benefits from medical treatment is a constantly changing process. If this book provides physicians and surgeons with at least some of the information they need to make such an assessment, it will have served its purpose.

The first chapter provides a brief overview of the concepts of risk, in the context of cancers arising after medical treatment. It is followed by chapters dealing with cancers following radiotherapy for both malignant and non-malignant disease, respectively, and with cancers following chemotherapy, also for malignant and non-malignant disease. Cancer risk following treatment for childhood malignancy is becoming an increasingly important topic, and this is treated in a separate chapter, which is followed by a review of cancer risks after various surgical procedures. The last chapter examines the various ways in which the risk of cancers caused by medical treatment may be reduced.

References

Penn, I. (ed.) (1982). Cancers induced by therapy. *Cancer Surveys*, **1,** 1–782.
Schmähl, D., Thomas, C., and Auer, R. (1977). *Iatrogenic carcinogenesis*. Springer, Berlin.

Acknowledgements

I would like to thank the many people who have helped in the preparation of this book. Frank Neal gave me the benefit of his immense experience in treating cancer patients. Calum Muir, Jimmy Elder, Henrik Møller, and especially Jacques Estève provided invaluable comments on various drafts. Eva Démaret gave very generously of her time in finding and checking references. Above all, Margot Geesink repeatedly typed and checked the entire book with unfailing good humour and the most awesome rigour. Any remaining errors are mine. I hope all these people will feel that their efforts were worthwhile.

Lyon
April 1991   M.P.C.

Contents

Contributors	xv

1 Risk and benefit — 1
SIMON SCHRAUB and MICHEL P. COLEMAN

Introduction	1
Risk	3
Risk and benefit	4
The future	8

2 Radiation treatment for cancer — 13
NICHOLAS E. DAY

Introduction	13
Radiotherapy for cervix cancer	13
Overall risk of second cancer	15
Breast cancer	19
Leukaemia	21
Myeloma	24
Other sites	24
Radiotherapy for Hodgkin's disease and ovarian cancer	26
Conclusion	27

3 Irradiation for non-malignant conditions — 29
SARAH C. DARBY

Introduction	29
X-ray treatment of ankylosing spondylitis	30
Women irradiated for benign gynaecological disorders	33
Patients treated with radium-224	36
Patients given Thorotrast	37
Irradiation for benign conditions in childhood	40
Irradiation of the breast	42
Patients treated with iodine-131	44
Conclusion	45

4 Cytotoxic chemotherapy for cancer — 50
JOHN M. KALDOR and CHRISTINE LASSET
- Introduction — 50
  - Sources of information — 52
- Leukaemia — 52
  - Leukaemia subtypes — 55
  - Temporal factors — 56
  - Leukaemogenic potency — 57
  - The influence of the first cancer — 60
  - Interaction of chemotherapy and radiotherapy — 61
- Bladder cancer — 61
- Other cancers — 63
- Prevention of chemotherapy-induced second cancer — 65

5 Chemotherapy and immunosuppression for non-malignant conditions — 71
MARK S. DORREEN and BARRY W. HANCOCK
- General introduction — 71
- Organ transplantation — 72
  - Aetiology — 73
  - Immunosurveillance — 73
  - Latency — 74
  - Viruses — 74
  - Specific cancers after organ transplantation — 76
  - Transplantation after cancer — 78
  - Prognosis — 78
- Malignancy in other benign diseases — 79
  - Lymphoma — 80
  - Acute myelogenous leukaemia — 83
  - Bladder cancer — 85
  - Kaposi's sarcoma — 86
  - Carcinoma of the uterine cervix — 88
  - Malignancy and the skin — 88
  - Other cancers — 89
- Conclusion — 90

6 Childhood malignancy — 101
JILLIAN M. BIRCH
- Introduction — 101
- Prenatal exposure — 104
  - *In utero* irradiation — 104
  - Exposure to drugs *in utero* — 106

Postnatal exposures for benign conditions	109
Radiation	109
Drugs	110
Immunization	110
Treatment for childhood malignancies	111
Overall risk of second malignancy	111
Radiation and chemotherapy	113
Cancer in the offspring of survivors	117
Genetic factors	117
Conclusion	121

7 Surgery 127
MICHEL P. COLEMAN

Introduction	127
Primary prevention	127
Partial gastrectomy	128
Cancer of the gastric remnant	128
Cancer at other sites	131
Cholecystectomy	133
Organs with immunological function	137
Splenectomy and thymectomy	137
Appendectomy and tonsillectomy	138
Vasectomy	140
Surgical implantation of a foreign body	141
Breast implants	142
Orthopaedic prostheses	144
Cardiac pacemakers	146
Cystoplasty and ureterosigmoidostomy	146

8 Reducing the risk 153
PATRICIA FRASER

Introduction	153
Population surveillance	154
Clinical surveillance	156
Hospital-based registries and hospital case series	156
Clinical trials	157
Biomonitoring	158
Modification of treatment regimes	161
Surgical procedures	161
Immunosuppressive therapy	162
Radiotherapy	162
Cytostatic therapy	163
Conclusion	169

Index 175

Contributors

Jillian M. Birch CRC Paediatric and Familial Cancer Research Group, Department of Epidemiology and Social Oncology, University of Manchester, Christie Hospital and Holt Radium Institute, Manchester M20 9BX, UK

Michel P. Coleman Unit of Descriptive Epidemiology, International Agency for Research on Cancer, 150, cours Albert Thomas, 69372 Lyon Cedex 08, France

Sarah C. Darby Cancer Epidemiology Unit, Imperial Cancer Research Fund, Gibson Building, Radcliffe Infirmary, Oxford OX2 6HE, UK

Nicholas E. Day MRC Biostatistics Unit, University of Cambridge, 5 Shaftesbury Road, Cambridge CB2 2BW, UK

Mark S. Dorreen Department of Clinical Oncology, University of Sheffield, Weston Park Hospital, Whitham Road, Sheffield S10 2SJ, UK

Patricia Fraser Department of Epidemiology and Population Sciences, London School of Hygiene and Tropical Medicine, Keppel Street, London WC1E 7HT, UK

Barry W. Hancock Department of Clinical Oncology, University of Sheffield, Weston Park Hospital, Whitham Road, Sheffield S10 2SJ, UK

John M. Kaldor Unit of Biostatistics Research and Informatics, International Agency for Research on Cancer, 150, cours Albert Thomas, 69372 Lyon Cedex 08, France
Current affiliation: National Centre in HIV Epidemiology and Clinical Research, University of New South Wales, Sydney, Australia

Christine Lasset Unité de Biostatistiques et d'Evaluation Thérapeutiques, Centre Léon Bérard, 18, rue Laënnec, 69008 Lyon, France

Simon Schraub Centre Hospitalier Régional, Hôpital Jean Minjoz, 1, boulevard Fleming, 25030 Besançon Cedex, France

1
Risk and benefit

SIMON SCHRAUB AND MICHEL P. COLEMAN

Introduction

Clinicians are increasingly aware of the risk of malignant disease arising as a complication of medical treatment, including treatment for cancer, but—as the following chapters of this book will make clear—information of adequate quality for assessing the size of this risk is generally recent. In some clinical settings the size of the risk, if any, is still uncertain. Treatment-related malignancies such as acute myeloid leukaemia and sarcoma are often particularly difficult to treat, however (Tucker *et al.* 1988; Robinson *et al.* 1988), and their importance can be judged by the broad changes in clinical policy which have already taken place in order to reduce the risk of their occurrence. Radiotherapy for functional or benign conditions such as mastitis, osteoarthritis and herpes zoster has been practically abandoned, for example, and 'mesna' is now routinely used to inactivate acrolein, a metabolite of the alkylating agent cyclophosphamide, presumed to be responsible for the carcinogenicity of cyclophosphamide to the bladder.

Selection of treatment regimes for individual patients, both for cancer and for other diseases, has also come to include an evaluation of the risks of subsequent cancer against the benefits of treatment. Thus alkylating agent therapy for rheumatological conditions such as systemic lupus and rheumatoid arthritis is reserved for the most intractable and aggressive cases, and adjuvant chemotherapy in Hodgkin's disease is increasingly avoided in stage I of the disease where there are favourable prognostic factors, for which radiotherapy alone can be expected to provide some 82 per cent disease-free survival at five years (Mauch 1989*a*).

The value of an overall assessment of the risk of long-term adverse effects of medical treatment will emerge clearly from the following chapters. Such an evaluation provides information which is essential if a reasonable balance of the risks and benefits of treatment is to be available to the clinician and—most important—explained to the patient. It also provides some reassurance that in many cases the cancer risks of current treatment are likely to be acceptable to the patients who are asked to run those risks. It further identifies a number of situations in which the balance of risk and benefit is far from obviously favourable. Finally, such an evaluation broadens our understanding of carcinogenesis, as with radiation-induced leukaemia and breast cancer (Day, Chapter 2, this volume), and the relative carcinogenic potency of different alkylating agents (Kaldor and Lasset, Chapter 4, this volume), and it provides a quantifiable rationale for the design of treatments which reduce the risk of subsequent cancer while maintaining or improving their efficacy in the management of the primary disease (Fraser, Chapter 8, this volume).

The gradual elucidation of various excess cancer risks related to medical treatment should also serve as a warning. Although radiation-induced cancer was recognized at the turn of the century, it took 50 years before there was a general recognition that malignancy could be caused by doses well below those causing serious acute effects (Darby, Chapter 3, this volume). Recognition has been swifter for the cancer risks associated with chemotherapy, but complex multi-stage regimes involving surgery, various modalities of radiotherapy, and mixtures of chemotherapy agents are now often being used in the search for improved treatment efficacy, particularly for cancer, and the possibility that such combinations may also incur an increased risk of subsequent cancer should not be overlooked. While the occurrence of acute toxicity is carefully monitored to assess the safety and short-term acceptability of such new treatment combinations, a more systematic approach is required for the detection of long-term treatment hazards such as cancer, and for precise estimation of the magnitude of any such risks. Ironically, this is particularly necessary for those treatments which are most successful in prolonging survival, and thus most widely used.

The benefits of treatment, particularly for cancer, are usually evaluated in terms of the cure rate, or an increase in the duration of disease-free survival. Measures of the patient's 'quality of life' after treatment are also being used in such assessments, but these measures still pose methodological and practical difficulties (De Haes and Knippenberg 1985; McMillen Moinpour *et al.* 1989; Donovan *et al.* 1989). The practising clinician is accustomed to drawing up, for each patient, a mental evaluation of the benefits of a recent treatment programme and its unwanted side-effects. Recent advances in the understanding of cancers related to treatment have altered this traditional evaluation in three ways.

First, the unwanted side-effects considered have come to include those that may arise many years later, in particular new malignancies. That this is possible at all is eloquent testimony to the benefits of treatment, since it is clear that 'a discussion of late effects and secondary malignancies is possible only if the primary treatment is successful' (Coleman and Tucker 1989). Second, such evaluations are frequently made in advance, such that the available knowledge of cancer risks related to a particular treatment regimen is incorporated into the balance of risk and benefit involved in choosing the treatment in the first place.

Inevitably, such a step incorporates a third element, that of probability. If we cannot know in advance which patient subjected to a particular treatment will develop a subsequent cancer, it may at least be possible to estimate the risk, or, more familiarly, to answer a question such as: 'If this patient is given that treatment, what are the chances that, in 10 years, he/ she will have survived without developing a malignant neoplasm?' This more quantitative approach is an advance over earlier clinical risk–benefit evaluations for at least two reasons: it should enable the choice of a less risky treatment with similar efficacy, and it gives the clinician the opportunity for a more rational discussion of the choice of treatment and any associated risks with the patient.

Risk

The risk of cancer following medical treatment can be estimated from several different study designs, and can be expressed quantitatively by various measures, the choice of which may affect the way in which the risk is perceived, both by the clinician and the patient. The study designs are outlined in Chapter 8 (Fraser). We will briefly consider the relevance and application of the various quantitative expressions of risk: fuller descriptions can be found in standard texts of epidemiology and statistics. For the sake of simplicity we will discuss them here in the context of the risk of a second cancer following the treatment of the first cancer, although the argument is equally applicable to cancer risk following treatment for non-malignant disease. We will also use the term 'cancer' somewhat loosely, to cover all malignant neoplasms including leukaemias, lymphomas, sarcomas etc.

The *percentage* of second cancers simply expresses the observed number of persons with a second cancer as a percentage of the number treated for the first cancer. Percentages are frequently used in reports of small case series, but they can be misleading and are often used incorrectly (see, for example, Coleman, Chapter 7, this volume). Unlike the actuarial risk (see below), a simple percentage takes no account of the duration of observation or survival of each patient in the series, and cannot easily be compared to cancer incidence in the general population in order to provide some idea of the risk relative to subjects who have not undergone the treatment of interest.

The *average annual incidence rate* relates the number of second cancers observed in a given period of time (the numerator of the rate) to the number of patients treated for a first cancer. Patients are observed for different periods of time, however, and should only be considered 'at risk' of a second cancer if they are still under observation and any second cancer is likely to be detected. Each patient's observation time is cumulated in the relevant time period (e.g. 0–9, 10–14 etc. years since the treatment) as person-years of observation (the denominator of the rate). Incidence rates calculated in this way are expressed per 100 000 person-years at risk, and may be applied either to the entire period of observation or to specific time periods since the treatment. Patients who die, emigrate, or are lost to follow-up cannot later contribute second cancers to the study (rate numerator) and do not therefore contribute person-time to the rate denominator after they cease to be eligible: the incidence rate thus takes precise account of the varying periods of follow-up.

The *relative rate* implies a contrast between the incidence rate observed in the treated group and the rate in a suitable comparison group. Ideally, this group would be similar to the treated group in all respects expect for the treatment, but the general population is often used, since cancer incidence (or mortality) rates will usually be available. The observed number of second cancers is compared with the number that would be expected if the treated group were to experience the same cancer risk as the comparison group in each category of sex, age, and calendar period (race, geographic region, etc.). The expected number of cancers is obtained as the product of the incidence rate in the comparison group (specific for each sex, age group, and calendar period) and the person-years at risk observed in the treated group in the same category of age, sex, and calendar period. The ratio of the total number of second cancers observed to the total of expected cancers derived in this way is an estimate of the ratio of incidence rates (adjusted for sex, age, and any trends over calendar time) between the study group and the comparison group; in other words, an estimate of the relative risk.

The *actuarial risk* provides the risk or probability of a second cancer over a given time period since treatment. It is calculated as a life-table probability from the follow-up details of the cohort of treated patients, and expressed as a percentage, such as a 1.1 per cent risk of leukaemia within 10 years of chemotherapy for breast cancer. The relative risk is the ratio between the risks in the treated group and the comparison group.

The *cumulative incidence rate* (Day 1987) over a given age range is sometimes used to express cancer risks following treatment. It is a close approximation to the actuarial risk, and has the important advantage that it is easy to calculate (it is the sum of the age-specific incidence rates over the age-range concerned) and since it is expressed as a percentage, it is readily interpreted.

Finally, the *odds ratio* is derived from case-control studies, which have been widely used to study cancer risk in relation to medical treatments. The odds ratio obtained from unbiased case-control studies provides a good estimate of the relative risk (Breslow and Day 1980).

Risk and benefit

The various different numerical expressions for cancer risk following medical treatment may lead to quite different interpretations of its clinical significance (Harris and Coleman 1989). Thus a relative risk of 10 for leukaemia following a particular treatment may be judged extremely high, and providing the study is well-designed, it certainly indicates strongly that the treatment causes leukaemia, but if the underlying risk is very small, an individual's risk of leukaemia following the treatment will still be small—even if 10 times larger than it would otherwise have been. Here, the absolute or actuarial risk over a given period may be more useful in balancing risk and benefit. Thus in the NSABP (National Surgical Adjuvant Breast and Bowel Project) study (Fisher et al. 1985), the relative risk of leukaemia following adjuvant chemotherapy for breast cancer was extremely high (RR = 24), but the crude percentage of patients in the trial who received chemotherapy and later developed leukaemia was only 0.36 per cent: this corresponds, however, to an actuarial risk at 10 years of 1.1 per cent. These very different figures represent different ways of quantifying the same phenomenon—cancer risk after medical treatment—and appreciation of the distinction is clearly relevant for clinicians discussing such risks with their patients. After discussion with her physician, for example, a patient may decide that the likely functional gain in the short term from alkylating agent therapy for severe rheumatoid arthritis is more important for her quality of life than the small associated risk of lymphoma in the more distant future. Conversely, the fact that the cancer risk from a given treatment is small may carry little weight if the patient takes the view that 'while the statistical risk of death ... is 0.1 per cent overall, to the individual it strikes that risk will appear to be 100 per cent' (Anon. 1990). Such judgements may lead a patient to refuse a treatment that is clearly beneficial.

Three examples may suffice to show that the benefits of treatment—at least in conventional terms of survival—often outweigh the risk of a subsequent malignancy.

For stage II B cervical cancer, for which surgery alone will usually be inadequate, radiotherapy is the treatment of choice, with a 5-year survival of around 65 per cent (Hoskins et al. 1989). The relative risk of second cancers following radiotherapy is about 1.1, relative to the general population, and slightly higher among 10-year survivors (Day, Chapter 2, this volume). This increase in cancer risk would probably be considered acceptably small by most patients with cervical cancer, given the survival advantage conferred by radiotherapy.

Again, 64 per cent of patients with stage III A Hodgkin's disease will survive 8 years with radiotherapy alone, and 55 per cent will be disease free, while these results are improved to practically 100 per cent and 94 per cent, respectively, by the addition of chemotherapy (Mauch, 1989b). But the risk of leukaemia following radiotherapy alone is increased more than 10-fold, while with adjuvant chemotherapy it is more than 100-fold (Tucker et al. 1988; Kaldor et al. 1990). The improvement in survival is obtained at a price—a much greater risk of leukaemia. Yet there is little doubt that this improvement must outweigh the 15-year actuarial risks of leukaemia (3.3 per cent) and solid tumours (13.2 per cent).

For multiple myeloma, the advantage is not so obvious. Chemotherapy protocols offer 30 per cent survival at 4 years (Bergsagel et al. 1979), but although the underlying risk of acute leukaemia in multiple myeloma patients is unknown, the actuarial risk of acute leukaemia probably attributable to the treatment is 17.4 per cent after the same period (50 months). The median survival in stage II and III myeloma before the availability of effective systemic chemotherapy was less than a year, and despite the very large risk of leukaemia, the overall prognosis of myeloma stages II and III has been considerably improved by chemotherapy, with a median survival in the range from 3 to 4 years (Salmon and Cassady 1989).

The objective of providing effective treatment while minimizing or avoiding any associated cancer risk remains of prime importance, even when the treatment is given for life-threatening malignancy. For stage I nephroblastoma with favourable histology, treatment without radiotherapy and with less aggressive chemotherapy regimes provides a good example, since post-operative radiotherapy and actinomycin are likely to be responsible for an excess risk of subsequent cancer (Breslow et al. 1988). Prophylactic cerebral irradiation in acute lymphoblastic leukaemia is no longer used because, in addition to serious intellectual and psychological disturbance (Meadows et al. 1981), it causes astrocytoma (Malone et al. 1986; Rimm et al. 1987). Similarly, for well-differentiated ovarian malignancies in stage IA or IB, a clinical trial has shown that chemotherapy regimes which include the alkylating agent melphalan, which is leukaemogenic, do not improve survival (Young et al. 1990).

A more complex situation arises with irradiation of the chest wall after mastectomy for breast cancer. This reduces the rate of local recurrence, but does not improve survival (Fisher et al. 1985; Host et al. 1986), yet it increases the risk of several types of cancer (Harvey and Brinton 1985). Thus localized sarcoma and angiosarcoma of the chest wall have been reported (Boice et al. 1985; Souba et al. 1986; Otis et al. 1986; Laskin et al. 1988; Kurtz et al. 1988), as have radiation-induced osteosarcomas (Doherty et al. 1986). The issue of leukaemia risk following radiotherapy of the chest wall remains unresolved: Boivin et al. (1986) noted a 2-fold risk in one study, but Curtis et al. (1989) concluded there was no increase in risk. Similarly, the increased risk of contralateral breast cancer due to irradiation observed in studies in Denmark (Storm and Jensen 1986), the UK (Haybittle et al. 1989) and the USA (Harvey and Brinton 1985), has not been detected in other cohort studies and trials (McCredie et al. 1975; Hankey et al. 1983; Fisher et al. 1985; Host et al. 1986). The evidence is certainly strong enough to argue that post-mastectomy irradiation should not be used routinely, or at least limited to particular cases where the clinical arguments are overwhelming.

Irradiation for benign tumours should also be abandoned. In the central nervous system, however, a difficult problem arises for localized tumours of low malignancy which are incompletely excised, since second malignant neoplasms of the brain have been observed after cerebral irradiation for both benign and malignant brain tumours (Anderson and Treip 1984; Liwnicz et al. 1985; Kumar et al. 1987; Cavin et al. 1990). There is nevertheless some evidence that irradiation may help both for incompletely resected benign meningioma (Taylor et al. 1988; Salazar 1988; Glaholm et al. 1990), and for incompletely resected low-grade astrocytoma, although the benefits of irradiation are less clear-cut in the latter situation (Piepmeier 1987; Whitton and Bloom 1990).

The potential late effects of treatments that are widely used for common tumours merit particular vigilance, simply because of the large numbers of patients exposed. Any additional risk of subsequent malignancy due to such treatments would give rise to large numbers of additional cancers. Following surgery for cancer of the female breast, for example, radiotherapy, chemotherapy, and adjuvant hormonal therapy have been used. Current conservative treatment for small tumours involves limited 'lumpectomy' and radiotherapy. Is there any long-term risk of a second cancer? So far, at least, there does not appear to be any increase in the risk of contralateral breast cancer up to 10 or 20 years after treatment (Kurtz et al. 1988; Veronesi et al. 1990). The results on acute leukaemia risk following adjuvant alkylating agent chemotherapy for breast cancer are contradictory: a 2.7-fold risk after cyclophosphamide in one study (Haas et al. 1987), but a 24-fold risk following melphalan in another (Fisher et al. 1985). This difference may reflect the relative carcinogenic potency of the two agents (Kaldor and Lasset, Chapter 4, this volume). In randomized trials, there appears as yet to be no increase in leukaemia risk after adjuvant chemotherapy for breast cancer (Bonadonna et al. 1985), but this treatment is increasingly widely used (Fisher et al. 1989), and in view of the known leukaemogenicity of alkylating agents there is a need for constant vigilance. The problem of hormonal therapy is very similar. Tamoxifen, an oestrogen antagonist with some agonist properties, is also widely used for breast cancer, particularly for tumours with positive oestrogen receptors. Cases of endometrial cancer have been described (Hardell 1988a,b; Fornander et al. 1989); these may be explained by the oestrogen-like activity of tamoxifen.

The issue of possible tamoxifen carcinogenicity is the more important in view of its widespread use for breast cancer, especially in premenopausal women, and the fact that it is even being proposed for use in the primary prevention of breast cancer (Fentiman 1990). This has two implications. First, the population exposed would be much larger than if restricted to patients with cancer, and the public health importance of any attached cancer risk would be commensurately greater. Second, evaluation of the balance of risk and benefit is quite different for a woman with a life-threatening cancer and a woman who is healthy: inevitably, the great majority of women using a drug for primary prevention of cancer would not develop the disease anyway, and the preventive efficacy of the drug will need to be known before a rational balance can be achieved. These considerations are likely to have an increasing impact on clinical practice in the future.

The future

The evidence reviewed in this book suggests to us that, both for the patient with serious disease and for the treating physician, increased duration of survival alone is no longer enough, and that disease-free survival may eventually come to mean survival free both of the original disease and of any new malignant disease complicating its treatment. If this is correct, then the responsibility of the clinical triallist examining the efficacy of new treatments should now be enlarged: major trial protocols will need to include a systematic evaluation of the risk of long-term side-effects, in addition to the treatment efficacy and short-term side-effects which are routinely studied today. The implications for the size, complexity, duration and cost of clinical trials are major (Byar 1988), but have not yet been seriously addressed.

One approach will be to combine the data from follow-up of patients in several trials; the advantages of so doing are discussed by Fraser (Chapter 8, this volume). The systematic approach adopted by the US National Cancer Institute (Greene 1984) to monitoring trial patients for late complications of alkylating agent therapy provides a model of what can and should be achieved. Its adoption elsewhere as a routine policy, whether by individual trial investigators or by a central organization with access to a national register of cancers or deaths (such as the UK National Health Service Central Registers) would considerably increase the chances of detecting and quantifying cancer risks from new medical treatments more reliably and rapidly in the future.

A patient who has a serious disease, requiring medical intervention of the type discussed in this volume, will usually be aware that he 'is often already at risk, or he would not have consulted us' (Anon. 1990). Such patients have also come to accept that there is always a risk associated with the treatment: not just the risk that it may not work, but the possibility that it may cause a new and perhaps even more serious risk to health. The challenge faced by physicians, surgeons and radiotherapists in their daily work is to achieve the right balance for each patient between the immediate risks of the disease and the longer-term risks associated with its treatment.

References

Anderson, J. R. and Treip, C. S. (1984). Radiation-induced intracranial neoplasm: a report of three possible cases. *Cancer*, **53**, 426–9.
Anonymous (1990). Ever so risky. *Lancet*, **336**, 216–17.
Bergsagel, D. E., Bailey, A. J., Langley, G. R., MacDonald, R. N., White, D. F., and Miller, A. B. (1979). The chemotherapy of plasma-cell myeloma and the incidence of acute leukemia. *New England Journal of Medicine*, **301**, 743–8.
Boice, J. D. Jr, *et al.* (ed.) (1985). *Multiple primary cancers in Connecticut and Denmark*, National Cancer Institute Monograph 68. National Institutes of Health, Bethesda.
Boivin, J.-F., Hutchison, G. B., Evans, F. B., Abou-Daoud, K. T., and Junod, B. (1986). Leukemia after radiotherapy for first primary cancers of various anatomic sites. *American Journal of Epidemiology*, **123**, 993–1003.
Bonadonna, G., Rossi, A., and Valagussa, P. (1985). Adjuvant CMF chemotherapy in operable breast cancer: ten years later. *Lancet*, **i**, 976–7.
Breslow, N. E. and Day, N. E. (1980). *Statistical methods in cancer research, Vol. 1, The analysis of case-control studies*, IARC Scientific Publications No. 32. International Agency for Research on Cancer, Lyon.
Breslow, N. E., Norkool, P. A., Olshan, A., Evans, A., and D'Angio, G. J. (1988). Second malignant neoplasms in survivors of Wilms' tumor: a report from the National Wilms' Tumor Study. *Journal of the National Cancer Institute*, **80**, 592–5.
Byar, D. P. (1988). The design of cancer prevention trials. In *Cancer clinical trials: a critical appraisal*, (ed. H. Scheurlen, R. Kay, and M. Baum), *Recent Results in Cancer Research*, **111**, 34–48.
Cavin, L. W., Dalrymple, G. V., McGuire, E. L., Maners, A. W., and Broadwater, J. R. (1990). CNS tumor induction by radiotherapy: a report of four new cases and estimate of dose required. *International Journal of Radiation: Oncology—Biology—Physics*, **18**, 399–406.
Coleman, C. N. and Tucker, M. A. (1989). Secondary cancers. In *Cancer: principles and practice of oncology*, (ed. V. T. DeVita, Jr, S. Hellman, and S. A. Rosenberg), 3rd edn, pp. 2181–90. Lippincott, Philadelphia.
Curtis, R. E., Boice, J. D. Jr, Stovall, M., Flannery, J. T., and Moloney, W. C. (1989). Leukemia risk following radiotherapy for breast cancer. *Journal of Clinical Oncology*, **7**, 21–9.
Day, N. (1987). Cumulative rate and cumulative risk. In *Cancer incidence in five continents*, Vol. V, IARC Scientific Publications No. 88, (ed. C. Muir, J. Waterhouse, T. Mack, J. Powell, and S. Whelan), pp. 787–9. International Agency for Research on Cancer, Lyon.
De Haes, J. and Knippenberg, F. (1985). The quality of life of cancer patients: a review of the literature. *Social Science and Medicine*, **20**, 809–17.
Doherty, M. A., Rodger, A., and Langlands, A. O. (1986). Sarcoma of bone following therapeutic irradiation for breast carcinoma. *International Journal of Radiation: Oncology—Biology—Physics*, **12**, 103–6.
Donovan, K., Sanson-Fisher, R. W., and Redman, S. (1989). Measuring quality of life in cancer patients. *Journal of Clinical Oncology*, **7**, 959–68.

Fentiman, I. S. (1990). Breast cancer prevention with tamoxifen: the role of tamoxifen in the prevention of breast cancer. *European Journal of Cancer*, **26**, 655–6.

Fisher, B., Rockette, H., Fisher, E. R., Wickerham, D. L., Redmond, C., and Brown, A. (1985). Leukemia in breast cancer patients following adjuvant chemotherapy or postoperative radiation: the NSABP experience. *Journal of Clinical Oncology*, **3**, 1640–58.

Fisher, B., *et al.* (1989). A randomized clinical trial evaluating sequential methotrexate and fluorouracil in the treatment of patients with node-negative breast cancer who have estrogen-receptor-negative tumors. *New England Journal of Medicine*, **320**, 473–8.

Fornander, T., *et al.* (1989). Adjuvant tamoxifen in early breast cancer: occurrence of new primary cancers. *Lancet*, **i**, 117–20.

Glaholm, J., Bloom, H. J. G., and Crow, J. H. (1990). The role of radiotherapy in the management of intracranial meningiomas: the Royal Marsden Hospital experience with 186 patients. *International Journal of Radiation: Oncology—Biology—Physics*, **18**, 755–61.

Greene, M. H. (1984). Interaction between radiotherapy and chemotherapy in human leukemogenesis. In *Radiation carcinogenesis: epidemiology and biological significance*, (ed. J. D. Boice, Jr and J. F. Fraumeni, Jr), pp. 199–210. Raven Press, New York.

Haas, J. F. *et al.* (1987). Risk of leukaemia in ovarian tumour and breast cancer patients following treatment by cyclophosphamide. *British Journal of Cancer*, **55**, 213–18.

Hankey, B. F., Curtis, R. E. Naughton, M. D., Boice, J. D. Jr, and Flannery, J. T. (1983). A retrospective cohort analysis of second breast cancer risk for primary breast cancer patients with an assessment of the effect of radiation therapy. *Journal of the National Cancer Institute*, **70**, 797–804.

Hardell, L. (1988*a*). Tamoxifen as risk factor for carcinoma of corpus uteri. *Lancet*, **ii**, 563.

Hardell, L. (1988*b*). Pelvic irradiation and tamoxifen as risk factors for carcinoma of corpus uteri. *Lancet*, **ii**, 1432.

Harris, J. R. and Coleman, C. N. (1989). Estimating the risk of second primary tumors following cancer treatment. *Journal of Clinical Oncology*, **7**, 5–6.

Harvey, E. B. and Brinton, L. A. (1985). Second cancer following cancer of the breast in Connecticut, 1935–82. In *Multiple primary cancers in Connecticut and Denmark*, (ed. J. D. Boice, Jr, *et al.*), National Cancer Institute Monograph 68, pp. 99–112. National Institutes of Health, Bethesda.

Haybittle, J. L., Brinkley, D., Houghton, J., A'Hern, R. P., and Baum M. (1989). Postoperative radiotherapy and late mortality: evidence from the Cancer Research Campaign trial for early breast cancer. *British Medical Journal*, **298**, 1611–14.

Hoskins, W. J., Perez, C., and Young, R. C. (1989). Gynecologic tumors. In *Cancer: principles and practice of oncology*, (ed. V. T. DeVita, Jr, S. Hellman, and S. A. Rosenberg), 3rd edn, pp. 1099–161. Lippincott, Philadelphia.

Host, H., Brennhovd, I. O., and Loeb, M. (1986). Postoperative radiotherapy in breast cancer—long-term results from the Oslo study. *International Journal of Radiation: Oncology—Biology—Physics*, **12**, 727–32.

Kaldor, J. M., et al. (1990). Leukaemia following Hodgkin's disease. *New England Journal of Medicine*, **322**, 7–13.

Kumar, P. P., Good, R. R., Skultety, F. M., Leibrock, L. G., and Severson, G. S. (1987). Radiation-induced neoplasms of the brain. *Cancer*, **59**, 1274–82.

Kurtz, M. J., Almaric, R., Brandone, H., Ayme, Y., and Spitalier, J. P. (1988). Contralateral breast cancer and other second malignancies in patients treated by breast-conserving therapy with radiation. *International Journal of Radiation: Oncology—Biology—Physics*, **15**, 277–84.

Laskin, W. B., Silverman, T. A., and Enzinger, F. M. (1988). Postradiation soft tissue sarcomas: an analysis of 53 cases. *Cancer*, **62**, 2330–40.

Liwnicz, B. H., Berger, T. S., Liwnicz, R. G., and Aron, B. S. (1985). Radiation-associated gliomas: a report of four cases and analysis of post-radiation tumors of the central nervous system. *Neurosurgery (Baltimore)*, **17**, 436–45.

Malone, M., Lumley, H., and Erdohazi, M. (1986). Astrocytoma as a second malignancy in patients with acute lymphoblastic leukemia. *Cancer*, **57**, 1979–85.

Mauch, P. M. (1989a). In Hellman, S., Jaffe, E. S., and DeVita, V. T., Hodgkin's disease. In *Cancer: principles and practice of oncology*, (ed. V. T. DeVita, Jr, S. Hellman, and S. A. Rosenberg), 3rd edn, p. 1712. Lippincott, Philadelphia.

Mauch, P. M. (1989b). In Hellman, S., Jaffe, E. S., and DeVita, V. T., Hodgkin's disease. In *Cancer: principles and practice of oncology*, (ed. V. T. DeVita, Jr. S. Hellman, and S. A. Rosenberg), 3rd edn, p. 1716. Lippincott, Philadelphia.

McCredie, J. A., Inch, W. R., and Alderson, M. (1975). Consecutive primary carcinomas of the breast. *Cancer*, **35**, 1472–7.

McMillen Moinpour, C., Feigl, P., Metch, B., Hayden, K. A., Meyskens, F. L., and Crowley, J. (1989). Quality of life end points in cancer clinical trials: review and recommendations. *Journal of the National Cancer Institute*, **81**, 485–95.

Meadows, A. T., Gordon, J., Massari, D. J., Littman, P., Fergusson, J., and Moss K. (1981). Declines in IQ scores and cognitive dysfunctions in children with acute lymphocytic leukaemia treated with cranial irradiation. *Lancet*, **ii**, 1015–18.

Otis, C. N., Peschel, R., McKhann, C., Merino, M. J., and Duray, P. H. (1986). The rapid onset of cutaneous angiosarcoma after radiotherapy for breast carcinoma. *Cancer*, **57**, 2130–4.

Piepmeier, J. M. (1987). Observations on the current treatment of low-grade astrocytic tumors of the cerebral hemispheres. *Journal of Neurosurgery*, **67**, 177–81.

Rimm, I. J., Li, F. C., Tarbell, N. J., Winston, K. R., and Sallan, S. E. (1987). Brain tumors after cranial irradiation for childhood acute lymphoblastic leukemia. *Cancer*, **59**, 1506–8.

Robinson, E., Neugut, A. I., and Wylie, P. (1988). Clinical aspects of postirradiation sarcomas. *Journal of the National Cancer Institute*, **80**, 233–40.

Salazar, O. M. (1988). Ensuring local control in meningiomas. *International Journal of Radiation: Oncology—Biology—Physics*, **15**, 501–4.

Salmon, S. E. and Cassady, J. R. (1989). Plasma cell neoplasms. In *Cancer: principles and practice of oncology*, (ed. V. T. DeVita, Jr, S. Hellman, and S. A. Rosenberg), 3rd edn, pp. 1853–87. Lippincott, Philadelphia.

Souba, W. W., McKenna, R. J., Meis, J., Benjamin, R., Raymond, A. K., and Mountain, C. F. (1986). Radiation-induced sarcomas of the chest wall. *Cancer*, **57**, 610–15.

Storm, H. H. and Jensen, O. M. (1986). Risk of contralateral breast cancer in Denmark 1943–80. *British Journal of Cancer*, **54**, 483–92.

Taylor, B. W., Marcus, R. B., Friedman, W. A., Ballinger, W. E., and Million, R. R. (1988). The meningioma controversy: postoperative radiation therapy. *International Journal of Radiation: Oncology—Biology—Physics*, **15**, 299–304.

Tucker, M. A., Coleman, C. N., Cox, R. S., Varghese, A., and Rosenberg, S. A. (1988). Risk of second cancers after treatment for Hodgkin's disease. *New England Journal of Medicine*, **318**, 76–81.

Veronesi, U., et al. (1990). Breast conservation is the treatment of choice in small breast cancer: long-term results of a randomized trial. *European Journal of Cancer*, **26**, 668–70.

Whitton, A. C. and Bloom, H. J. C. (1990). Low grade glioma of the cerebral hemispheres in adults: a retrospective analysis of 88 cases. *International Journal of Radiation: Oncology—Biology—Physics*, **18**, 783–6.

Young, R. C., et al. (1990). Adjuvant therapy in stage I and stage II epithelial ovarian cancer: results of two prospective randomized trials. *New England Journal of Medicine*, **322**, 1021–7.

2
Radiation treatment for cancer

NICHOLAS E. DAY

Introduction

People given radiotherapy for cancer constitute the largest group given high doses of radiation under controlled circumstances. They might therefore be expected to yield much useful information on the quantitative relationship between the characteristics of the exposure and the subsequent risk of cancer. Since some malignancies are treated by a combination of radiotherapy and chemotherapy, one might also anticipate extensive information on the risk associated with joint exposure, with well-determined dose-response relationships. There are, however, serious problems in the interpretation of much of the data relating to second primary cancers, and these problems have greatly limited the contribution such data might have been expected to make, both to our understanding of radiation carcinogenesis and to the assessment of the hazards of low levels of radiation.

By far the largest study of radiation-induced second cancers has been of women treated for cancer of the cervix. It is instructive to examine this study in detail, because it exemplifies both the value and the limitations of such studies.

Radiotherapy for cervix cancer

Cancer of the cervix has been treated since the early years of the century by the introduction of radium implants into the vagina and uterine cavity, where they are left for two to three days. These treatments (intracavitary treatment) were later supplemented by various types of external beam radiotherapy. Radium has in recent years been supplanted by other radioactive sources. With intracavitary treatment (brachytherapy), something approaching a point source of radiation is accurately implanted in the body for a length of time which is accurately known, and measurement of the radiation dose received by the other organs in the body can therefore be made, by simulating typical treatments on a life-size model of the human body, or phantom, such as the Alderson Rando phantom (Stovall 1983). For external beam therapy (teletherapy), a wide range of machines and fields have been used over the years, but similar phantom simulation is also feasible. For the purposes of estimating the cancer risk from radiation directly in humans, study of cervical cancer patients provides both the advantage of being able to estimate accurately the dose received at different points in the body and the availability for study of a very large exposed population. Many tens of thousands of women have been treated for cancer of the cervix in Europe and North America over the past 60 years, often in centres where long-term follow-up of cancer patients has been routinely practised. The situation was therefore highly favourable for epidemiological assessment of the risk for second cancers that the radiation treatment for cancer of the cervix had induced. Most of this chapter is devoted to a description of a large international study of cervix cancer patients.

The initial study was based on a number of cancer treatment centres in western Europe and the US. It recruited about 30 000 women currently undergoing treatment in the early 1960s and followed them actively for five years. The endpoints of interest were limited to leukaemia and its precursor blood abnormalities. The hope was to obtain further information on the radiation–leukaemia relationship, to add to the information available from the studies of A-bomb survivors (Kato and Schull 1982) and of patients irradiated for ankylosing spondylitis (Smith and Doll 1982). On the basis of these two studies, well over a hundred leukaemia cases were anticipated in the cervical cancer study. Only 13 leukaemias were observed, however, even less than the 15.5 leukaemias that would have been expected from a comparable population of women who had not been irradiated (Boice and Hutchison 1980). This paradoxical result was initially uninterpretable. Several years elapsed before experimental results in mice showed that the dose-response for radiation-induced myeloid leukaemia was not linear, nor even steadily increasing with dose (monotonic). At high doses, the leukaemia rate started to decrease with increasing radiation dose. This reversal of the slope of the dose-response was attributed to a cell-killing effect, such that above a certain dose of radiation the cells that were susceptible to malignant transformation were killed rather than transformed. Leukaemia was thus less likely to develop.

In the early 1980s, the study of cervical cancer patients was restarted and greatly enlarged. Many cancer registries in Europe and North America follow the individuals they have registered for life, also registering any subsequent malignancy that may occur. They therefore provide an excellent mechanism for the identification and follow-up of large cohorts of patients with particular malignancies. Retrospective search through the records of the institutions where they were treated enables subsequent risk of a second malignancy to be related to the treatment the individual received for a first cancer. Collaborating cancer registries identified nearly 100 000 patients with invasive cancer of the cervix, of whom 82 616 had a record of receiving radiotherapy (Day and Boice 1983). Together with the original cancer centres and further extensions, a total of over 150 000 women with invasive cervix cancer was assembled. Case-control studies were then performed of specific second malignancies within this cohort, which provide most of the dose-response data from the material. Initially, however, a straightforward cohort study was undertaken, using the cancer registry data (Day and Boice 1983). This study used only two-thirds of the total cohort and only had rudimentary treatment details, but because the external comparison group (the general population) of unirradiated persons was very large, some radiation-related conclusions emerge more clearly than from comparisons within the cohort. The cohort approach also allows the absolute excess of a second malignancy at any site to be estimated, which is not possible with conventional case-control studies.

To provide a perspective from which to view the cohort study results, Table 2.1 shows the main types of treatment which have been given for cervix cancer over the years, and Table 2.2 gives the typical radiation dose received by different organs in the body under these treatment regimes (Stovall 1983). As expected, very high doses are received by most organs in the pelvic region, doses of 1 to 3 Gy (100–300 rad) to the kidney, pancreas, and stomach, between 0.1 and 0.5 Gy to the lungs and breasts, and lower doses to the thyroid and other organs in the head and neck.

Overall risk of second cancer

Among the 82 616 women for whom cancer registry data indicated that they had received radiotherapy, the observed and expected second malignancies are shown in Table 2.3. There is an overall excess of cancer of 8.5 per cent (3324 observed versus 3062.5 expected). If attention is confined to the period 10 years or more after cervix cancer diagnosis, the overall excess is 11.9 per cent. Almost the entire excess can be accounted for by malignancies at sites close to the cervix, which received 50 Gy (5000 rad) or more (see Table 2.3). Among all other sites taken together, there is no overall excess. There is, however, a wide range of excesses and deficits at individual sites, most of which are related to factors other than radiation to the site in question. These factors include the misclassification of metastases, the lack of comparability between cervix cancer patients and the general population with respect to risk factors for other cancers, and non-carcinogenic effects of irradiation. For some sites, several factors may play a role. The large excess of lung cancer is an example. Cervix cancer patients smoke more than the general population, and this excess of smoking would be expected to generate about a 2-fold excess of lung cancer. The overall excess (Table 2.3) is 3.6-fold, but this single figure hides as much as it reveals. Table 2.4 breaks down this excess by age at diagnosis and time since diagnosis of the cervix cancer. Very high relative

Table 2.1 Commonly used systems of radiotherapy for cancer of the uterine cervix

	Stockholm system		Manchester system		Fletcher system	
	External beam[1]	Intracavitary[2]	External beam[1]	Intracavitary[2]	External beam[1]	Intracavitary[2]
Stage I	None	7000	None	9800	None	10 000
Stage II	30 (^{60}Co) or 20 (Ortho)	7000	None	9800	30 (Betatron) or 35 (Ortho)	7500
Stage III	40 (^{60}Co) or 24 (Ortho)	7000	30 (^{60}Co) or 25 (Ortho)	8500	55 (Betatron) or 55 (Ortho)	5000
Stage IV	45 (^{60}Co) or 24 (Ortho)	None	50 (^{60}Co)	3000	70 (Betatron) or 60 (Ortho)	None

[1] Dose (Gy) to midline pelvis.
[2] In mgh.
Data from Stovall (1983).

Table 2.2 Mean dose (Gy) to other body organs from typical radiotherapy for cancer of the cervix: by stage of cervix cancer

		Stage I	Stage II	Stage III	Stage IV
Radiotherapy regime					
Intracavitary (radium)[1]		9000	8000	7000	–
External beam[2]					
250 kVp X-rays			30	40	60
^{60}Co		–	30	40	50
25 MV X-rays		–	40	50	70
Bladder	250 kVp	54.00 +	78.00 +	82.00 +	60.00
	^{60}Co	54.00 +	78.00 +	82.00 +	50.00
	25 MV	54.00 +	88.00 +	92.00 +	70.00
Rectum	250 kVp	54.00 +	78.00 +	82.00 +	60.00
	^{60}Co	54.00 +	78.00 +	82.00 +	50.00
	25 MV	54.00 +	88.00 +	92.00 +	70.00
Ovaries	250 kVp	14.00	33.70	39.20	42.60
	^{60}Co	14.00	43.00	51.60	51.00
	25 MV	14.00	53.60	62.40	72.10
Kidneys	250 kVp	1.99	2.94	3.43	3.76
	^{60}Co	1.99	1.80	1.90	1.22
	25 MV	1.99	1.59	1.59	0.93
Stomach	250 kVp	0.99	2.55	3.00	3.34
	^{60}Co	0.99	1.57	1.69	1.15
	25 MV	0.99	1.40	1.42	0.91
Pancreas	250 kVp	1.13	2.47	2.84	2.95
	^{60}Co	1.13	1.61	1.68	1.01
	25 MV	1.13	1.37	1.34	0.64
Lungs	250 kVp	0.19	0.35	0.39	0.37
	^{60}Co	0.19	0.35	0.39	0.31
	25 MV	0.19	0.30	0.32	0.24
Breasts	250 kVp	0.15	0.33	0.37	0.38
	^{60}Co	0.15	0.33	0.37	0.32
	25 MV	0.15	0.28	0.29	0.25
Thyroid	250 kVp	0.06	0.13	0.15	0.15
	^{60}Co	0.06	0.15	0.17	0.16
	25 MV	0.06	0.16	0.17	0.18
Salivary glands	250 kVp	0.04	0.09	0.10	0.11
	^{60}Co	0.04	0.09	0.11	0.10
	25 MV	0.04	0.12	0.14	0.15
Brain	250 kVp	0.03	0.08	0.09	0.11
	^{60}Co	0.03	0.06	0.07	0.06
	25 MV	0.03	0.11	0.13	0.15
Active bone marrow					
Total	^{60}Co	3.25	9.24	11.00	10.60
Total, excl. pelvis	^{60}Co	1.00	2.68	3.17	3.00

[1] mgh.
[2] Absorbed dose at midline on central axis, expressed in Gy.
Data from Stovall (1983).

Table 2.3 Observed (Obs) and expected (Exp) numbers[1] of second primary cancers after radiotherapy for invasive cancer of the cervix

Second primary cancer	(ICD7)	Obs	Exp	Obs/Exp
Buccal cavity and nasopharynx	(140–148)	60	46.61	1.3*
Oesophagus	(150)	40	27.32	1.5*
Stomach	(151)	200	210.37	1.0
Small intestine	(152)	21	9.46	2.2**
Colon	(153)	314	301.53	1.0
Rectum	(154)	197	157.38	1.3**
Liver	(155.0)	19	19.88	1.0
Gallbladder	(155.1)	45	55.69	0.8
Pancreas	(157)	120	95.88	1.3*
Nose	(160)	13	7.12	1.8
Larynx	(161)	16	7.21	2.2**
Lung	(162–163)	491	134.94	3.6***
Breast	(170)	569	804.41	0.7***
Corpus uteri	(172)	126	209.13	0.6***
Other uterus	(173)	21	18.11	1.2
Ovary	(175)	136	198.31	0.7***
Other genital	(176)	101	43.00	2.4***
Kidney	(180)	69	66.50	1.0
Bladder	(181)	194	73.74	2.6***
Melanoma	(190)	36	47.08	0.8
Other skin	(191)	206	219.76	0.9
Eye	(192)	7	8.47	0.8
Brain	(193)	51	67.54	0.8*
Thyroid	(194)	36	31.47	1.1
Bone	(196)	11	5.72	1.9
Connective tissue	(197)	27	14.57	1.9***
Lymphoma	(200,202)	61	53.77	1.1
Hodgkin's disease	(201)	14	16.87	0.8
Multiple myeloma	(203)	33	33.92	1.0
All leukaemia	(204)	77	65.83	1.2
Chronic and unspecified lymphatic leukaemia	(204.0)	18	22.64	0.8
Myeloid and acute leukaemia	(204.1–4)	58	40.94	1.4*
Sites close to the cervix[2]		1601	1479.27	1.1***
Total (all sites except cervix)		3324	3062.54	1.1***

*$0.01 < p < 0.05$; **$0.001 < p < 0.01$; ***$p < 0.001$.
[1] Numbers exclude the first year of observation.
[2] Stomach, small intestine, colon, rectum, liver, gallbladder, pancreas, corpus uteri, other uterus, ovary, other genital, kidney, bladder, bone, and connective tissue.
Data from Day et al. (1983).

Table 2.4 Lung cancer among women irradiated for invasive cancer of the cervix: number of cases observed (Obs) and relative risk (RR), by age at diagnosis and time since diagnosis of cervix cancer

Age at diagnosis of cervix cancer	Time since diagnosis of cervix cancer (years)					
	1–4		5–9		10+	
	Obs	RR[1]	Obs	RR	Obs	RR
< 40	24	23.5	16	7.4	10	1.1
40–49	49	8.0	58	6.6	44	2.1
50–59	52	4.2	63	4.8	57	3.1
60+	51	2.6	45	3.2	24	2.5
Total	176	4.5	182	4.8	135	2.3

[1] Ratio of observed to expected numbers; expected numbers calculated from age–sex–period-specific lung cancer rates within each contributing registry.

risks (up to 20-fold) are seen in the first 5 years of follow-up, particularly for younger women. Beyond 10 years of follow-up the excess is little more than would be predicted from the smoking association. Another explanation must be found for the excess in the first 10 years of follow-up, and the simplest is that some or most of these apparent lung cancers are misclassified metastases from the cervix cancer; squamous epithelial tumours are the most frequent carcinoma at each site, and cervix cancer is known to metastasize to the lung. Given the large effects of smoking and of misclassification of metastases, any moderate increase in the risk of lung cancer (e.g. relative risk of 1.3 or so) due to radiation would be difficult to detect. Since the radiation dose to the lung from typical radiotherapy for cervical cancer is in the range of 0.1 to 0.5 Gy, no more than a moderate increase in risk would be expected, so the cohort study is basically uninformative with regard to lung cancer. Smoking also seems to provide a plausible explanation for the observed excess of cancers of the oral cavity, oesophagus, pancreas, and larynx.

Breast cancer

The deficit of breast cancer in this cohort illustrates the fact that radiation may influence cancer incidence by mechanisms other than direct carcinogenesis. The breakdown of breast cancer occurrence by time and age is given in Table 2.5. There is a large deficit among women who were irradiated when under 40 years of age, and this deficit increases with time since irradiation. Since the radiotherapy for cervical cancer would have destroyed ovarian function, these women in fact underwent a radiation-

Table 2.5 Breast cancer among women irradiated for cancer of the cervix: observed numbers (Obs) and relative risks (RR) by age at diagnosis and time since diagnosis of cervix cancer

Age at diagnosis of cervix cancer	Time since diagnosis of cervix cancer (years)											
	1–4		5–9		10–14		15–19		20+		Total	
	Obs	RR[1]	Obs	RR	Obs	RR	Obs	RR	Obs	RR	Obs	RR
<40	9	0.48	9	0.30	15	0.49	5	0.25	3	0.18	41	0.35
40–49	42	0.59	62	0.84	35	0.67	26	0.89	17	0.71	182	0.73
50–59	59	0.77	51	0.74	46	1.01	14	0.59	11	0.73	181	0.79
60+	66	0.69	68	1.05	22	0.67	8	0.67	2	0.47	166	0.79

[1] See footnote to Table 2.4.

induced menopause. The data in this table are by far the clearest and most extensive on the long-term effect on breast cancer risk of eliminating ovarian function below the age of 40. For women aged over 50 when irradiated, there is a 20 per cent deficit of breast cancer, which is roughly constant with increasing time since irradiation. This deficit might be attributed to the earlier pregnancies and higher parity of cervix cancer patients, but direct calculation suggests that this reproductive profile would only generate a 10 per cent deficit. Ten per cent is similar to the deficit seen in non-irradiated cervix cancer patients (data not shown). Even in women aged over 50, therefore, radiation to the ovaries appears to have a small protective effect on breast cancer risk.

The remaining major discrepancies between observed and expected second cancers are the large apparent deficits seen for endometrial and ovarian cancer: these organs were often removed at surgery for the cervical cancer, however. Minor excesses are seen for tumours of the small intestine, bone and connective tissue. An excess was seen of acute and non-lymphocytic leukaemia, especially in the first 10 years (45 observed, 24.2 expected), and of multiple myeloma, especially 15 years or more after the cervix cancer diagnosis. These two observations will be considered in more detail below, with the results from the extended case-control study.

The main findings from this cohort study are clear. Major doses of radiation to the site of the cervix cancer lead to an overall increase of some ten per cent in the total future cancer burden for the women concerned, and most of this radiogenic excess of cancer is limited to neighbouring organs which received doses from 10–90 Gy. When individual cancer sites are considered there were many differences between the numbers of observed second cancers and the numbers expected from general population rates, but most of these differences were not due to the carcinogenic effect of radiation.

Leukaemia

A more detailed picture of radiation-induced malignancy emerges when the case-control results are considered. The results for leukaemia are of special interest (Boice *et al.* 1987). From the extended cohort of 150 000 women with cervical cancer, 195 leukaemias were identified, of which 51 were chronic lymphoid leukaemia (CLL). Induction of CLL was not related to radiation in this series, nor has it been in other series, so the cases of CLL are excluded from further consideration. An attempt was made to match four controls to each case. After exclusion of cases and controls for whom no good quality radiation exposure data were available, there were 134 cases of acute leukaemia and chronic myeloid leukaemia, and 534 controls. In order to consider dose-response relationships, attention has to be paid to the heterogeneity of the dose to different compartments of the bone marrow. The distribution of radiation dose to the scattered elements of bone marrow from the radiotherapy protocols used for cervical cancer is given in Table 2.6. The heterogeneity is extreme: only 10 per cent of the bone marrow receives a radiation dose in the restricted range of radiation (1–5 Gy; 100–500 rad) for which leukaemia risk is known to be high.

Table 2.6 Distribution of active bone marrow and average dose per anatomic site from radiotherapy for cervical cancer[1]

Anatomic site	Per cent of active bone marrow	Dose to bone marrow (Gy)		
		Brachytherapy only	External beam only	All radiotherapy
Humeri	2.3	0.08	0.18	0.15
Clavicles	0.8	0.05	0.12	0.09
Femurs, top quarter[2]	3.4	3.80	4.70	5.80
Femurs, second quarter[2]	3.4	1.10	1.82	1.90
Pelvis and sacrum	27.4	7.90	27.80	20.20
Ribs and sternum	19.2	0.31	0.35	0.41
Scapulae	2.8	0.13	0.17	0.18
Cranium	7.6	0.02	0.06	0.05
Mandible	0.8	0.02	0.08	0.05
Lumbar vertebrae 1 and 2	4.6	0.95	2.77	2.40
Lumbar vertebrae 3 and 4	5.2	2.30	10.00	7.60
Lumbar vertebra 5	2.5	5.00	38.50	25.10
Thoracic spine	16.1	0.31	0.44	0.47
Cervical spine	3.9	0.02	0.08	0.05
All bone marrow	100.0			

[1] These doses are derived from dosimetry estimates for all cases and controls combined.
[2] The femur was divided into four quarters. In the age range at which cervical cancer occurs, the lower half of the femur contains no active marrow.
Data from Boice *et al.* (1987).

Risk models for non-uniform marrow irradiation Given this uneven distribution of radiation to the bone marrow, radiation-leukaemia dose-response models can perhaps best be obtained by summing the components of risk over all the various compartments of bone marrow. Thus, if

(1) marrow compartment 'i' represents a proportion w_i of the total bone marrow; and
(2) this marrow compartment receives a radiation dose d_i; and
(3) the relative risk of leukaemia arising somewhere in the marrow from a dose 'd', uniformly applied to all the marrow, can be represented as a function of dose, $f(d)$;

then the overall dose-response for leukaemia from the irregularly distributed radiation dose received by the various components of bone marrow during radiotherapy for cervical leukaemia can be modelled as a simple weighted average of the risks from each marrow component, using the proportion of the total marrow (w_i) in each component as the weight:

$$\text{overall relative risk} = \sum_i w_i \, f(d_i).$$

Of course this expression for the relative risk reduces again to $f(d)$ if all 'i' components of marrow receive the same dose; it expresses the idea that if all parts of the marrow are equally sensitive to radiation leukaemogenesis, then for a given dose of radiation, the risk of leukaemia arising in a given anatomical component of marrow depends only on the proportion of the total marrow it contains. Since the dose actually varies between marrow components, the expression given above is the simplest way to model the leukaemia risk.

Several choices can be made for the way in which risk is actually assumed to vary with dose, i.e. the risk function ($f(d)$). The risk function may be flat (the 'null' hypothesis that radiation has no effect on leukaemia risk); or linear (risk increases steadily with dose); or linear with cell killing (risk declines above a certain dose); or quadratic (accelerating risk increase with dose), with or without cell killing.

The main results from this analysis are given in Table 2.7, and Table 2.8 provides a summary of the dose-response analyses that have been conducted. It is claimed that 'both the linear-exponential and the quadratic-exponential models provided reasonable fits to the data ... (but) both the linear and the exponential models provided poor fits and could be rejected as unsuitable' (Boice *et al.* 1987), but this claim is not supported by the authors' own analysis. None of the models differ significantly from the 'null' model, under which the relative risk of leukaemia is completely independent of radiation dose (Table 2.8, column 1). Even the comparison of ever versus never radiotherapy is only just significant, with a one-sided test at the 5 per cent level (data not shown).

Table 2.7 Relative risk of acute leukaemia and chronic myeloid leukaemia, following radiation therapy for cervix cancer

Dose category (Gy)	Average bone marrow dose (Gy)	Numbers Cases	Numbers Controls	RR matched	90% CI
0	0	8	56	1.0	
0.01–2.49	1.90	10	48	1.37	0.5–3.5
2.50–4.99	3.60	27	80	2.53	1.1–6.0
5.00–7.49	6.30	36	128	2.11	0.9–4.8
7.50–9.99	8.60	34	130	1.97	0.9–4.3
10.00–12.49	11	13	62	1.62	0.7–4.0
12.50–14.99	13	2	8	1.50	0.5–4.4
≥ 15	17	4	22	1.42	0.5–4.1

The original publication of the case-control study results took no account of the fact that, in the cohort component of the study, there was a highly significant excess of acute and non-lymphocytic leukaemia in the first 10 years after irradiation (45 observed, 24.2 expected, RR 1.80, 90 per cent CI 1.43–2.40). The problem is the sparsity of cases with no radiation exposure (Table 2.7), thus making uncertain the baseline from which relative risks are calculated in the conventional case-control analysis. A more complete analysis of the data would incorporate a comparison with the underlying risk in the general population (i.e. the expected number of leukaemia cases in the total cohort), as well as the temporal evolution of

Table 2.8 Goodness of fit of different dose-response models from the international radiation study of women with cervical cancer: improvement of fit over the null model

Dose-response model	DF	Analysis based on: Case-control study alone[1] Likelihood ratio X^2	p-value	Case-control and cohort data combined[2] Likelihood ratio X^2	p-value
Linear	1	0.70	0.403	6.76	0.01
Linear with cell killing	2	4.10	0.129	10.44	0.005
Linear with cell killing, with latency	3	–	–	16.95	0.001

[1] Boice et al. (1987).
[2] Thomas et al. (1991).

leukaemia risk following exposure. Incorporating this extra information into the analysis would considerably sharpen the inferences that could be made about the shape of the dose-response. Such analysis have now been performed (Thomas et al. 1991). They show that an excellent fit to the observed data is provided by a model which incorporates a linear dose-response with a cell-killing effect at higher doses, and terms for different periods of latency, each of these parameters providing a highly significant improvement in fit over the baseline model (Table 2.8, column 2).

Combining the data from this study with data from a study in which the dose distribution over the bone marrow is very different, for example a Hodgkin's disease series, might markedly narrow the range of consistent models, and produce a much clearer insight into the biology of radiation carcinogenesis.

Myeloma

Multiple myeloma is another example where the very small number of non-exposed individuals within the cohort rendered the case-control results, taken on their own, almost meaningless. Several studies have shown an excess of myeloma 15 years or more after radiation exposure. The cohort phase of this study provided strong confirmation of such an excess, confined to this same period of follow-up, from a comparison with the general population: 15 years or more after the initial diagnosis of cervix cancer, 16 multiple myeloma cases were observed among the 82 616 women given radiotherapy, against 7.58 expected ($p < 0.005$, one-sided). There were, in contrast, 17 cases in years 1–14 of follow-up against 26.33 expected, and the difference between the two periods is highly significant ($X^2 = 13.0, p < 0.001$). In the case-control study, only two cases and two controls were not exposed, and all the confidence intervals for the estimated relative risks were wide, both for sub-groups defined by dose and those defined by time since exposure.

Other sites

The most interesting results from the overall study are probably the negative ones. For cancers of the stomach, pancreas, and kidney, organs which received average radiation doses of about two Gy, no significant excess was seen among irradiated women in the initial cohort study, and no significant dose-response was seen in the case-control study (Table 2.9). When the results for the stomach are seen in conjunction with the results from the A-bomb survivors and the ankylosing spondylitis series, a strong indication of a quadratic dose-response curve emerges (Day 1985). For cancer of the colon, which typically receives a dose that varies from 3 to 30 Gy along its length during radiotherapy for cervical cancer, no excess is seen at all.

Table 2.9 Dose-response data for cancers of the stomach, pancreas and kidney among women irradiated for cancer of the cervix and surviving at least 10 years

Dose category (Gy)	Stomach			Pancreas			Kidney		
	Cases	Controls	RR	Cases	Controls	RR	Cases	Controls	RR
0	6	21	1.00	8	16	1.00	6	25	1.00
Up to 1	20	42	1.79	20	35	1.09	6	12	1.82
1–1.99	93	170	2.04	27	53	0.82	23	38	2.38
2–2.99	90	174	1.99	50	82	1.15	25	46	2.30
≥ 3	27	42	2.39	14	33	0.69	8	17	1.71
One-sided p-value for trend			0.12			0.37			0.17

Data from Boice et al. (1988).

Only for cancers of the rectum, bladder and other sites in the genital tract is there any evidence of a dose-response relationship, and this only at doses of over 50 Gy (5000 rad).

For cancer of the breast, the lack of apparent carcinogenic effect of 0.5 Gy (50 rad) is reassuring for women who undergo low-dose mammography. This series is by far the largest, in terms of the number of breast cancer cases, for dose levels in the 0 to 0.5 Gy range, and a radiogenic effect is notably absent (Table 2.10).

This group of studies of cervix cancer patients illustrates well the complementary advantages of the cohort and the case-control approaches. While case-control studies can provide accurate individual estimates of radiation dose, cohort studies provide a much larger unexposed comparison group. If the results from both approaches can be taken together, one achieves a more coherent interpretation of the results—equivalent, one might say, to binocular vision. Analyses using only the case-control data view the data through only one eye.

Table 2.10 Breast cancer following cancer of the cervix: relationship of risk to organ dose among 10-year survivors

Dose category (Gy)	Average dose to breast (Gy)	Cases	Controls	RR
0	0	71	132	1.0
up to 0.24	0.17	172	345	0.86
0.25–0.49	0.33	269	527	0.94
≥ 0.5	0.59	40	83	0.93

Data from Boice et al. (1988).

Radiotherapy for Hodgkin's disease and ovarian cancer

Two other groups of cancer patients which have been studied, taking similar advantage of the opportunities to assemble large cohorts of individuals through cancer registries, are survivors of ovarian cancer and of Hodgkin's disease (Kaldor *et al.* 1987). The contrast with the cervix cancer patients is striking. For both Hodgkin's disease and ovarian cancer, chemotherapy and radiotherapy were often used, either alone or in combination. In the ovarian cancer series, a large number of patients were treated solely by surgery. The overall excess of acute leukaemia is large in both series, with a 17-fold excess among patients with Hodgkin's disease and a 4-fold excess among the ovarian cancer patients (Kaldor *et al.* 1987). The excess is very much greater in those undergoing chemotherapy alone, rather than radiotherapy alone, and there is no evidence of a major synergistic effect on leukaemia induction by the two modalities combined. Some results are given in Table 2.11 (Kaldor *et al.* 1990a, b). Clear dose-response relationships were seen for each of the alkylating agents used commonly in the past for treatment of ovarian cancer, the relative risk for the higher dose levels reaching 20-fold or more. For Hodgkin's disease, the effectiveness of combination chemotherapy has limited the availability of information about individual agents, but increases in risk are evident for increasing numbers of cycles of the commoner treatment combinations. It is interesting that the 17-fold leukaemia risk for the higher doses of radiotherapy given for Hodgkin's disease is appreciably higher than the radiation-associated risk of leukaemia in cervix cancer patients. The distribution of dose over the bone marrow is markedly different for typical radiotherapy in the two series; a combined analysis might well shed considerable light on the underlying dose-response per unit of marrow.

Table 2.11 Relative risk of acute and non-lymphocytic leukaemia following chemotherapy and/or radiotherapy for Hodgkin's disease and ovarian cancer

Treatment	Hodgkin's disease RR (95 per cent CI)	Ovarian cancer RR (95 per cent CI)
Surgery alone	–	1.0
Radiotherapy alone	1.0^1	1.6 (0.51–4.8)
Chemotherapy alone	9.0 (4.1–20.0)	12.0 (4.4–32.0)
Radiotherapy and chemotherapy combined	7.7 (3.9–15.0)	9.8 (3.4–28.0)

[1] Radiotherapy alone is taken as the baseline risk in Hodgkin's disease, since very few cases were treated with neither radiotherapy nor chemotherapy.
Data from Kaldor *et al.* (1990a,b).

The long-term sequelae of treatment for childhood cancer have also been subject to large-scale multi-centre studies. A major excess of osteosarcomas following radiation therapy for retinoblastoma has been observed (Kingston *et al.* 1987; Mike *et al.* 1982). This subject is discussed in detail by Birch (Chapter 6, this volume).

Conclusion

The two great advantages of studying radiation-induced cancer through follow-up of treated cancer patients are that the organ doses for each patient can be accurately determined, and that there is the potential for amassing very large cohorts. There are, however, features of such studies which severely limit the applicability of their conclusions to the general population. First, the distribution of dose to the body is far from uniform, in particular being very large to the target organ. This virtually rules out the possibility of useful inferences about the dose-response relationship for tissues which are distributed throughout the body, particularly bone marrow, but also connective tissue, skin, and bone. It also renders many of the conclusions irrelevant for the major social and environmental problems involving radiation, which relate mainly to low doses at low dose-rates, i.e. extended over time. Second, cancer patients are different in many respects from the general population, and these differences may interfere with or even vitiate any general conclusions, as is the case with lung and ovarian cancer (to name but two) following cervix cancer. One major conclusion can be drawn from such studies, however: radiation exposure of the type given in cancer radiotherapy is only weakly carcinogenic, even at the highest doses which tissue can withstand.

References

Boice, J. D. and Hutchison, G. B. (1980). Leukaemia in women following radiotherapy for cervical cancer. Ten-year follow-up of an international study. *Journal of the National Cancer Institute*, **65**, 115–29.

Boice, J. D. Jr, *et al.* (1987). Radiation dose and leukaemia risk in patients treated for cancer of the cervix. *Journal of the National Cancer Institute*, **79**, 1295–311.

Boice, J. D. Jr, *et al.* (1988). Radiation dose and second cancer risk in patients treated for cancer of the cervix. *Radiation Research*, **116**, 3–55.

Day, N. E. (1985). Statistical considerations. In *Interpretation of negative epidemiological evidence for carcinogenicity*, IARC Scientific Publications No. 65, (ed. N. J. Wald and R. Doll), pp. 13–27. International Agency for Research on Cancer, Lyon.

Day, N. E. and Boice, J. D. Jr (ed.) (1983). *Second cancer in relation to radiation treatment for cervical cancer*, IARC Scientific Publications No. 52. International Agency for Research on Cancer, Lyon.

Day, N. E., et al. (1983). Summary chapter. In *Second cancer in relation to radiation treatment for cervical cancer*, IARC Scientific Publications No. 52, (ed. N. E. Day and J. D. Boice, Jr), pp. 137–81. International Agency for Research on Cancer, Lyon.

Kaldor, J. M., et al. (1987). Second malignancies following testicular cancer, ovarian cancer and Hodgkin's disease: an international collaborative study among cancer registries of the long-term effects of therapy. *International Journal of Cancer*, **39**, 571–85.

Kaldor, J. M., et al. (1990a). Leukemia following chemotherapy for ovarian cancer. *New England Journal of Medicine*, **322**, 1–6.

Kaldor, J. M., et al. (1990b). Leukemia following Hodgkin's disease. *New England Journal of Medicine*, **322**, 7–13.

Kato, H. and Schull, W. J. (1982). Studies of the mortality of A-bomb survivors. 7. Mortality, 1950–1978: Part I. Cancer mortality. *Radiation Research*, **90**, 395–432.

Kingston, J. E., Hawkins, M. M., Draper, G. J., Marsden, H. B., and Kinnier Wilson, L. M. (1987). Patterns of multiple primary tumours in patients treated for cancer during childhood. *British Journal of Cancer*, **56**, 331–8.

Mike, V., Meadows, A. T., and D'Angio, G. J. (1982). Incidence of second malignant neoplasms in children: results of an international study. *Lancet*, **ii**, 1326.

Smith, P.G. and Doll, R. (1982). Mortality among patients with ankylosing spondylitis after a single treatment course with X rays. *British Medical Journal*, **284**, 449–460.

Stovall, M. (1983). Organ doses from radiotherapy of cancer of the uterine cervix. In *Second cancer in relation to radiation treatment for cervical cancer*, IARC Scientific Publications No. 52, (ed. N. E. Day and J. D. Boice, Jr), pp. 131–6. International Agency for Research on Cancer, Lyon.

Thomas, D. C., Blettner, M., and Day, N. E. (1991). Use of external rates in nested case control studies with application to the International Radiation Study of Cervical Cancer Patients. *Biometrics* (in press).

3
Irradiation for non-malignant conditions

SARAH C. DARBY

Introduction

The first man-made radiation exposures arose from medical radiology. Following the discovery of X-rays in 1895 the value of radiographs was quickly and widely recognized, and by the 1930s the use of radiation had been extended to treat a wide range of diseases and conditions. At about this time, it was discovered that radioactive forms of essentially all the common elements could be produced artificially, and the subsequent development of nuclear reactors enabled the production of substantial quantities of man-made radionuclides for use in medicine as well as in industry and research. For a fuller historical account of the discovery and uses of radiation, see Pochin (1983).

It was very soon recognized that exposure to radiation could have harmful consequences; reddening and inflammation of the skin from artificially produced X-rays was reported in 1896, and by 1902 it had been noticed that cancer sometimes developed in frequently or heavily irradiated areas of skin. After that it did not take long to realize that cancer might develop in almost any body organ or tissue that had been damaged and scarred by large or repeated doses of irradiation. However, it was not until the mid-1950s that it was fully realized that exposure to ionizing radiation at doses insufficient to result in gross tissue damage could still be the cause of leukaemia and other cancers, and that these radiation-induced cancers would be indistinguishable clinically from those due to other causes.

There has never been any systematic attempt to quantify the full toll of radiation-induced cancer that has arisen following irradiation for non-malignant conditions. A considerable number of individual groups of irradiated patients have, however, been studied and the present chapter attempts to summarize the major findings in these groups. In many cases the therapy delivered to these groups is no longer used, and the main reason for continuing to study them is their suitability for quantifying the risks of cancer to populations likely to be exposed in the future to low doses of ionizing radiation, such as those exposed occupationally in the nuclear industry and elsewhere, and populations screened for disease using procedures that involve some radiation exposure.

X-Ray treatment of ankylosing spondylitis

Court-Brown and Doll identified over 14 000 patients with ankylosing spondylitis who had been treated with X-irradiation at some time between 1935 and 1954 at any one of 87 radiotherapy centres in Great Britain and Northern Ireland. Recent estimates based on the original treatment records of a 1 in 15 sample of patients indicate the mean total body dose in the group to be about 1.9 Gy, and that the great majority of individual organs in the trunk received doses in excess of 1 Gy (Lewis *et al.* 1988). When the patients were followed up, a clear excess in mortality was seen, both of leukaemia (Court-Brown and Doll 1957) and of other malignant neoplasms (Court-Brown and Doll 1965). However, these initial reports included many patients who had been treated with X-rays for their spondylitis more than once, and it was not clear whether the increase in cancer risk, which persisted for many years, should be attributed to the first course or subsequent courses. In more recent reports this difficulty has been avoided by restricting analysis to patients who received only a single course of treatment, and retreated patients have been retained in the study for only 18 months after their second course of treatment (Smith and Doll 1982; Darby *et al.* 1987).[1]

By 1 January 1983, the date of the most recent follow-up, a total of 3121 patients had died, that is just under 50 per cent of those who only received one course of treatment. Table 3.1 compares the numbers of deaths from various causes with the numbers expected if the patients in this study suffered the same age-, sex-, and time-specific death rates as the population of England and Wales. The number of deaths from neoplasms was 33 per cent greater, and that from other causes 51 per cent greater than expected. A group of spondylitics who were not irradiated has also been studied (Radford *et al.* 1977; Smith *et al.* 1977) and it has been shown that they also have a similar increase in mortality for causes other than cancer when comparison is made with the general population, while for cancer the number of observed deaths was almost exactly equal to that expected using national rates. It seems likely, therefore, that in the irradiated group the excess of deaths from causes other than cancer is a consequence of their spondylitis, while the excess of cancer is a consequence of their radiotherapy, with the possible exception of colon cancer, which may be associated with spondylitis through the increased risk of ulcerative colitis suffered by these patients (Hickling and Wright 1983). On this basis, over 150 radiation-induced deaths have occurred in the group.

The relative risk of mortality from leukaemia subdivided by time since

---

[1] Retreated patients were retained in the study for 18 months after retreatment because some patients may have been retreated for symptoms attributable to cancer that were misdiagnosed as spondylitis.

Table 3.1 Observed and expected deaths and relative risks up to age 85 in irradiated spondylitics, by major groups of disease

Cause of death	Observed deaths (Obs)	Expected deaths (Exp)	Relative risk (Obs/Exp)
All causes	3121***	2133.41	1.46
All neoplasms	727***	547.21	1.33
Leukaemia	39***	12.29	3.17
Colon cancer	47*	36.11	1.30
Other neoplasms	639***	498.76	1.28
All other causes[1]	2394***	1586.20	1.51

*$p<0.05$; ***$p<0.001$ (one-sided tests).
[1] Includes three deaths for which the cause was unknown.

first treatment is shown in Fig. 3.1. The highest relative risk was 12.5, observed in the period 2.5–4.9 years after treatment. There was a significant ($p < 0.001$) decline in the relative risk after this period, but more than 15.0 years after treatment the relative risk was still high at 1.87, and still statistically significant ($p < 0.05$). There is no evidence from these data that the increased leukaemia risk has disappeared completely, nor that it was changing materially more than 15 years after exposure. The long

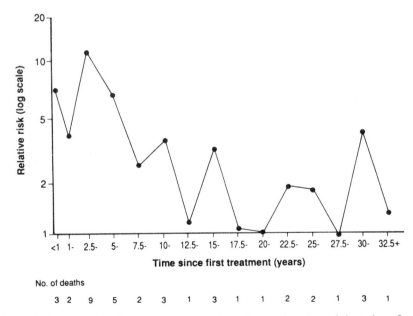

Fig. 3.1 Relative risk of mortality from leukaemia as a function of time since first treatment.

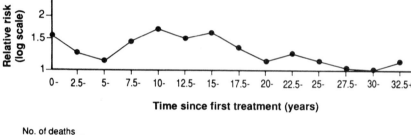

Fig. 3.2 Relative risk of mortality from neoplasms other than leukaemia or colon cancer as a function of time since first treatment.

duration of the risk is in agreement with results from recent studies of survivors of the atomic bombings of Hiroshima and Nagasaki, in which the relative risk of leukaemia more than 30 years after exposure is still around 2-fold (Darby *et al.* 1985).

The relative risk of mortality from all neoplasms other than leukaemia or colon cancer, subdivided by time since first treatment, is shown in Fig. 3.2. The relative risk was high (1.57) in the first 2.5 years, but fell to just over 1.1 in the period 5.0–7.4 years after treatment. Some of the tumours presenting in the period soon after treatment may have caused symptoms that were incorrectly ascribed to spondylitis, and thus some of this early excess of cancer may indirectly have 'caused' the radiotherapy, rather than have been caused by it. More than 7.5 years after treatment, the relative risk increased again, and it remained greater than 1.5 between 10 and 17.4 years after treatment. More than 17.5 years after treatment the relative risk declined, and the decreasing trend was highly significant ($p < 0.001$). From 25.0 years after treatment the number of observed deaths was only slightly greater than the number expected (178 against 166.56) and for this period the relative risk was 1.07 (95 per cent CI 0.92–1.24).

This is the first large study to suggest an apparent end to the effects of exposure to external radiation for neoplasms other than leukaemia. It remains to be seen whether the patterns of radiation-induced risk in other studies, such as those carried out of survivors of the atomic bombings of Hiroshima and Nagasaki, and of patients irradiated for cancer of the cervix, will follow suit. Further analyses of the spondylitis data have, however, produced no indication that the findings are spurious. For example, there is evidence that patients who remained in the study for 25 years or more received, on average, doses similar to those of patients included at earlier time periods (Lewis *et al.* 1988), and an analysis of trends in mortality from tobacco and alcohol related diseases, including ischaemic heart disease, chronic obstructive lung disease and cirrhosis of the liver

(Darby and Doll 1988) shows that there is no overall decline in relative risk from these diseases in the later period of the study. This tends to rule out the possibility that the irradiated spondylitics altered their lifestyle in such a way as to reduce their risk of cancer relative to the general population and thus provides no evidence that the use of national mortality rates might produce an inappropriate estimate of the number of cancer deaths that would have been expected in the absence of radiotherapy.

Women irradiated for benign gynaecological disorders

Irradiation was first given as a treatment for benign uterine bleeding in the early years of this century. Its use became widespread, and in many centres it remained popular until the 1960s. Treatment could be either by means of radium implants or by external X-ray therapy. The objective of the treatment was often to produce an artificial menopause, although in some centres there was a policy of giving only sufficient radiation to permit a return to normal menstruation.

The possibility that the irradiated women were more likely to develop subsequent pelvic malignancy was examined in a number of studies as early as the 1940s. Many of these early studies included only a limited follow-up period, however, and the first report in which an excess was demonstrated appeared in 1956 (Palmer and Spratt 1956). These authors mailed questionnaires to 1670 patients treated at Roswell Park Memorial Institute in New York State between 1930 and 1940 and, basing their analyses only on the 721 patients from whom adequate replies were received, found a greater than 3-fold increase when comparison was made with incidence data for New York State (61 cases observed compared with 17.40 expected).

During the 1950s, papers began to appear documenting the increased death rate from leukaemia among the survivors of the atomic bombings of Hiroshima and Nagasaki who had received whole body uniform irradiation, consisting chiefly of gamma rays in a single dose. This radiation-related excess of leukaemia was soon confirmed in other groups, such as the patients treated with X-irradiation for ankylosing spondylitis already described. In 1960, however, a study was published of 71 000 women treated for carcinoma of the cervix with intracavitary radium implants, with or without additional X-irradiation, in which the risk of developing leukaemia did not appear to be increased (Simon *et al.* 1960; see also Day, Chapter 2, this volume). This apparent anomaly led several investigators to initiate studies of women given pelvic irradiation for benign gynaecological diseases. Like the women with cervical cancer, these women received irradiation to a relatively small part of the bone marrow, mainly in the pelvis, lumbar spine and upper femur, but the mean marrow dose was substantially less than that given for carcinoma of the cervix. The results of two of these studies are summarized below.

One cohort included 2067 women with metropathia haemorrhagica who were given X-irradiation in three Scottish radiotherapy centres between 1940 and 1960 to produce an artificial menopause (Doll and Smith 1968; Smith and Doll 1976). By the time of the latest published follow-up, 500 (24.2 per cent) of the women had died. They were followed up from the date of their radiotherapy for an average of 19.0 years. Compared to the number of deaths expected based on Scottish mortality rates, there was a 2.6-fold increase in deaths from leukaemia, and a 29 per cent increase in deaths from cancers of sites judged to have been heavily irradiated (ovaries, intestines, rectum, uterus, other pelvic organs, and bladder); both increases were statistically significant (Table 3.2). Despite the large increases in leukaemia and cancers of heavily irradiated sites, the relative risk for all

Table 3.2 Numbers of deaths observed and expected and relative risk in a cohort of 2067 women given radiation-induced menopause in Scotland and followed for an average of 19.0 years

Cause of death	Observed deaths (Obs)	Expected deaths (Exp)	Relative risk (Obs/Exp)
Leukaemia	7	2.69	2.60*
Cancer of heavily irradiated sites	61	47.28	1.29*
Lymphoma and multiple meyloma	7	4.41	1.59
Breast cancer	14	26.86	0.52**
Other cancers	57	54.83	1.04
All neoplasms	146	136.07	1.07
All other causes	354	363.06	0.98
All causes	500	499.13	1.00

*$p<0.05$; **$p<0.01$ (one-sided tests).

neoplasms in this group is only 1.07, due mainly to a large deficit of deaths from cancer of the breast (Table 3.2). This deficit cannot be entirely due to the induction of an unusually early menopause, as it extended to women aged over 50 at the time of irradiation, and its cause is not yet understood.

The mean bone marrow dose for the women with metropathia, averaged over the entire body, has been estimated to be 1.34 Gy; if a simple linear dose-response relationship is assumed, the excess leukaemia death rate was 1.22 per 10^4 person-year Gy (95 per cent CI 0.13–3.13). When the women were subdivided into three dose groups the excess death rate tended to increase with increasing dose, but the numbers were too small to allow a firm conclusion. At the time of publication the excess death rate per Gy was in line with the excess seen among the Japanese atomic bomb survivors. The authors concluded that the likely explanation for the apparent discrepancy between the results of this study and those of the studies of women irradiated for cervix cancer, who had received much larger doses, lay in a cell killing effect in the heavily irradiated pelvic marrow of the cervix cancer patients. Subsequent studies of other cohorts of women with cervix cancer have confirmed this interpretation. Revised estimates of the excess death rate from leukaemia in the atomic bomb survivors, however, now indicate that the excess in that group is just over 3 per 10^4 person-year Gy (Preston and Pierce 1988). This raises the possibility that there may also have been some cell death in the irradiated bone marrow of the metropathia patients. Further follow-up of these women is under way.

Wagoner (1969, 1984) studied 1893 women with a variety of benign

Table 3.3 Numbers of primary cancers observed and expected and relative risk among women receiving radiotherapy for benign gynaecological disorders, Connecticut, 1935–1966

Site of cancer	Observed deaths (Obs)	Expected deaths (Exp)	Relative risk (Obs/Exp)
Leukaemia	12	5.3	2.3*
Female genital organs	109	54.9	2.0**
Urinary organs	17	8.5	2.0**
Lymphatic tissue and multiple myeloma	15	6.5	2.3**
Breast	54	57.9	0.9
Other and unknown sites[1]	103	96.9	1.06
Total	310	230.0	1.3**

*$p<0.05$; **$p<0.01$.
[1] Excluding non-melanoma skin cancer.

gynaecological disorders treated by X-ray or radium implants in five hospitals in Connecticut during the period 1935–1964. The mean bone marrow dose to these patients was estimated to be 1–3 Gy for the X-ray treated group, and 0.4–1.3 Gy for the radium treated group. When the numbers of incident cancers in the group were compared with the numbers expected based on Connecticut Tumor Registry rates there were 2-fold increases of leukaemia and cancers of female genital and urinary organs, in accord with the results from the metropathia cohort (Table 3.3). There was, however, no large deficit of cancers of the breast. Twelve cases of uterine sarcoma were reported, in contrast to only 1.5 expected. In the women treated with radium, the radium was implanted all around the endometrial wall, and when women treated with radium were examined separately from those treated by X-rays, the excess of sarcomas was seen to be concentrated in the radium group. Most were found to be mixed mesodermal carcinosarcomas, in contrast with the Connecticut registry data for the general population, in which approximately three quarters of uterine sarcomas reported were leiomyosarcomas. There is thus strong evidence that most of these sarcomas were induced by the radium implant.

Patients treated with ^{224}radium

After the Second World War, a series of about 2000 children and adults in Germany were treated for tuberculosis, ankylosing spondylitis and some other diseases with repeated intravenous injections of ^{224}radium (Mays 1973). Radium is chemically similar to calcium, and bone is the principal repository for radium which enters the body. ^{224}Radium emits alpha particles, and has a half-life of only 3.6 days, so almost all the radiation dose is delivered to the bone surface, before the radium is incorporated more widely throughout the bone volume. Other important alpha-emitting radionuclides such as plutonium are also initially concentrated by various mechanisms on the bone surface (Vaughan 1986), and it is for this reason that such a careful study has been made of patients treated with ^{224}radium.

It has not proved possible to follow up the entire group of 2000 patients, but from 1952 a subgroup of 899 patients has been studied. Results have been published for follow-up to June 1984, by which time half the patients were known to have died, and the average length of follow-up was 22 years (range 0–38 years) (Mays et al. 1986). The average injected dose of ^{224}radium was 18 μCuries/kg leading to an average skeletal dose of just over 4 Gy. However, those aged under 20 at first injection received on average nearly twice the injected dose of those aged at least 20 at first injection and, because of the enhanced uptake of radium by growing bone, their estimated skeletal dose was about 5-fold that of adults (10.6 Gy compared with 2.1 Gy). The time during which the dose was delivered ranged from 1 to 45 months, with a mean of 11 months.

An increased incidence of leukaemia (five cases observed versus two expected), renal cancer (four cases observed versus one expected), and a range of non-malignant diseases including benign osteochondromas has been observed in these patients. However, the most notable feature of their subsequent health to date is that they have developed 53 bone sarcomas. Cancer incidence rates for the German Democratic Republic give an expected number of only 0.2 bone sarcomas, and there is no evidence that either ankylosing spondylitis or tuberculosis increase the incidence of bone cancer in non-irradiated patients. Thus all, or nearly all, these bone sarcomas are likely to have been radiation-induced.

The first bone sarcoma appeared only 3.5 years after injection, the peak frequency after 6–8 years, and the last sarcoma after 25 years. No new tumours have appeared in the last 10 years of follow-up and the temporal pattern of expression was similar for those first irradiated in childhood and as adults. Thus it seems likely that most, if not all, of the radiation-induced bone sarcomas have already appeared in this group. The average cumulative risk of a bone sarcoma per Gy per person has been estimated to be about 0.017, and the risks per Gy appear roughly equal among men and women, among those irradiated for tuberculosis and spondylitis and, somewhat surprisingly, among those who were aged less than 20 when irradiated and older people (Mays et al. 1986). Examination of the dose-response relationship indicates that a simple linear model does not fit the observed data very well, and that it is necessary also to include a quadratic term in the model. When this is done the cumulative risk per Gy per person at low doses is estimated to be 0.0085, that is about half the risk indicated by a simple linear calculation (Chmelevsky et al. 1986). There is also some evidence of a tendency towards an increased risk per Gy if the period over which injections were given was increased: since the half-life of ^{224}radium is only 3.6 days, this would have caused the period of irradiation of bone to be prolonged (Mays and Spiess 1984).

It was shown by Spiess that ^{224}radium was ineffective in the treatment of tuberculosis (Mays 1973). For ankylosing spondylitis, however, the treatment continued to be thought to have therapeutic value, and although the original high dose treatment was discontinued in Germany after about 1950, a method of treatment involving much lower doses of ^{224}radium continued to be used, and is still used in several places. In order to monitor the present ^{224}radium therapy of ankylosing spondylitis, a series of 1501 adult patients treated mainly between 1948 and 1975 has been followed up (Wick and Gössner 1983; Wick et al. 1986). The majority of the patients received one series of 10 weekly injections of 28 µCuries of ^{224}radium, although some patients received further series of injections, and for the total group of patients the resulting mean skeletal dose is estimated at 0.65 Gy. By July 1984, the date of the most recent published results, the average length of follow-up was about 16 years, and ranged up to 34 years. By this time one bone tumour (a fibrosarcoma of the ilio-sacral joint) had been observed compared with about 0.5 expected from national rates. Thus, the large risk of osteosarcoma associated with the original high dose therapy seems to have been avoided so far in the patients studied after treatment with lower doses.

Patients given Thorotrast

Thorotrast was a commercially prepared radio-opaque agent used to improve image contrast in medical radiography, consisting of a 25 per cent colloidal solution of thorium dioxide. It was used in several countries from about 1930 until after the Second World War, usually administered as an intravascular injection. It was used extensively for cerebral angiography, although it was also used for other purposes and administered by other routes. After injection of Thorotrast, thorium is retained in the body and deposited at various sites in approximately the following proportions: liver 59 per cent, spleen 29 per cent, red bone marrow 9 per cent, calcified bone 2 per cent, lungs 0.7 per cent and kidneys 0.1 per cent (Kaul and Noffz 1978). Thorium has an extremely long physical half-life (more than 10 billion years), and decays by emitting an alpha particle and creating a series of radioactive daughters, including ^{224}radium. Persons with a body burden of thorium are therefore being continuously irradiated by ^{224}radium and its alpha-emitting daughters. For further details, see National Research Council (1988).

Five epidemiological studies of Thorotrast-exposed patients have been carried out in Germany (van Kaick et al. 1984, 1986), Japan (Mori et al. 1983, 1986), Portugal (Horta et al. 1978), Denmark (Faber 1983, 1986), and the USA (Falk et al. 1979). The most extensively documented is the German study, which was started in 1968. Just over 5000 patients known to have received Thorotrast were identified, as was a control group of similar size drawn from the same hospitals and with a similar age and sex distribution. In order to reduce the influence of the underlying diseases in the patients, any Thorotrast-exposed patients and controls known to have died within 3 years of injection or hospitalization were excluded from further study. Among the remainder the level of follow-up was poor, with just over 1100 Thorotrast-exposed patients and about 2700 controls untraced, chiefly due to the disruption caused by the Second World War. Nevertheless, the results based on the remaining 2334 exposed and 1912 control patients who could be followed up until 1984 are striking. Among the Thorotrast-exposed patients there have been 347 deaths from liver cancer, compared with only two in the controls (Table 3.4). The majority of these were carcinomas, primarily of the cholangiocellular type, and haemangiosarcomas. The shortest interval between injection of Thorotrast and death from liver cancer was 16 years, and it seems likely that the full burden of disease in the group has not yet become apparent, since a substantial number of liver cancer deaths have occurred in the most recent period of follow-up, with intervals of more than 40 years between injection and subsequent death being observed. The estimated typical dose of alpha radiation to the liver of these patients is about 0.25 Gy per year, with estimates of accumulated doses to individuals ranging from 2 to 15 Gy. In addition to the increase in liver cancer, there also appears to be a substantial increase in deaths from myeloproliferative disorders (mainly acute myeloid leukaemia) in the exposed group, with 35 deaths observed in Thorotrast patients and only three among the controls. There are also

Table 3.4 Numbers of deaths by cause in exposed and control patients: German Thorotrast Study

Cause of death	Exposed to Thorotrast	Controls
Liver cancer	347	2
Myeloproliferative disease	35	3
Chronic lymphatic leukaemia	3	2
Non-Hodgkin's lymphoma	16	7
Bone sarcoma	4	1
Lung cancer	46	40
Pleural mesothelioma	4	0
Kidney cancer	4	2
Liver cirrhosis	292	42
Bone marrow failure	20	1
Cardiovascular diseases	587	468
Other causes	606	841
Total number of deaths	1964	1409
Total number of patients with follow-up	2334	1912

increases in bone sarcomas, with four deaths observed among the exposed, and one among the controls, and in pleural mesothelioma with four deaths observed in the exposed, compared with none in the controls. There may also be proportionally smaller increases of other cancers, and there is a huge increase in deaths from liver cirrhosis, although in common with other studies of the effects of radiation, there is little evidence of an increase in chronic lymphatic leukaemia. Results of the studies carried out on patients given Thorotrast in other countries are similar.

In view of the enormous increase in liver cancers observed in these patients, compared with only a small increase in bone sarcomas, it is at first sight surprising that there was no excess of liver cancer in the patients treated with ^{224}radium described above. The explanation for the apparent discrepancy lies in the fact that, due to its physical colloidal form, much of the Thorotrast deposited in the liver becomes encapsulated in fibrous tissue there, so that only a small proportion of the ^{224}radium and its other daughter products can escape (Vaughan 1986). The different experience of these two groups of patients illustrates the complexities of the dosimetry of radionuclides, and demonstrates the difficulty of extrapolating from the effect of an ingested radionuclide in one situation to its effect in another, even when the routes of administration are superficially somewhat similar.

Irradiation for benign conditions in childhood

Until the long-term carcinogenic risks were appreciated, external radiotherapy involving irradiation of the head and neck region was used to treat several diseases and conditions of childhood, including ringworm of the scalp (*tinea capitis*), presumed enlargement of the thymus, whooping cough, and even various conditions of the tonsils and adenoids. Large numbers of individuals were involved. For example, in the Chicago area alone it is believed that over 70 000 patients were treated in the 1940s and 1950s (Shore *et al.* 1986). The chief harmful effect of such treatment appears to be an increased risk of thyroid cancer, although increases in skin and breast cancers, and also brain tumours have been observed (Shore *et al.* 1984; Hildreth *et al.* 1985; Ron and Modan 1984).

Radiation-induced thyroid cancer is difficult to study for a number of reasons. First, the principal cell type of cancer induced by ionizing radiation is papillary, which has a low case fatality rate (National Research Council 1980), and so mortality data are unsatisfactory. Second, incidence data are subject to bias in that screening and examination for thyroid cancer will detect substantial numbers of asymptomatic cancers. Thus any quantification of the risks of radiation-induced thyroid cancer involves carrying out incidence studies that include a control group of non-exposed subjects in whom the level of surveillance for thyroid cancer is similar to that in the exposed group.

One study that comes close to satisfying these requirements is the study including 2652 persons who received X-ray treatment in infancy for an 'enlarged' thymus in Rochester, New York, and 4823 sibling controls, who have been followed up for at least 5 years (Shore *et al.* 1985). By the date of the most recently published survey, the average length of follow-up was 30 years and 30 thyroid cancers had been detected in the exposed group, compared with only one among the controls, giving a relative risk of 49.1 after adjustment for sex, ethnicity, and interval since irradiation (90 per cent CI 10.7–225). The excess rate per Gy was slightly higher in the period 15–24 years after irradiation than at earlier or later periods, but there was still an excess risk in the later part of the follow-up more than 35 years after irradiation. In addition to these thyroid cancers there was also a large excess of surgically removed benign thyroid adenomas with 59 cases in the irradiated group against only eight among controls, giving an adjusted relative risk of 15.0 (90 per cent CI 8.1–27.7).

One of the strengths of this study is the wide range of thyroid doses received by the exposed individuals. Thyroid doses have been estimated from the original radiotherapy records and range from 0.05 to about 11 Gy. When thyroid cancer incidence was examined for various levels of estimated thyroid dose, the risk was estimated to be 3.5 per 10^4 person-year

Irradiation for non-malignant conditions

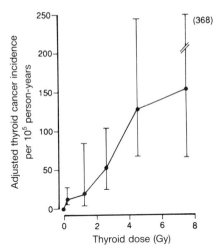

Fig. 3.3 Adjusted thyroid cancer incidence in relation to thyroid radiation dose, with 90 per cent confidence limits. Adjustment has been made for sex, ethnicity, and interval since irradiation.

Gy after adjustment for sex, ethnicity, and interval since irradiation. The dose-response relation was approximately linear, with little evidence of either curvature or cell-killing, even in the highest dose category (Fig. 3.3). When the dose-response analysis was carried out separately for males and females the risks were higher for females (5.25 per 10^4 person-year Gy) than for males (2.05 per 10^4 person-year Gy) and the difference was highly statistically significant. In most populations not specifically exposed to radiation, thyroid cancer rates in females are also higher than those in males, and the parallel sex difference in excess risk in the exposed group suggests that the radiation-induced risk tends to multiply the underlying risk of thyroid cancer, rather than add to it.

The largest study of children treated with X-irradiation to induce epilation (hair-loss) for ringworm of the scalp has been carried out in Israel (Ron and Modan 1980, 1984). It included 10 842 irradiated subjects and a control group made up of 5400 non-irradiated siblings and 10 842 population controls. About half of both study and control groups were male. Data on cancer incidence were obtained from the Israel Cancer Registry and searches of hospital records. By the time of the latest published findings the average length of follow-up was 23 years, and 29 thyroid cancers had been identified in the irradiated subjects compared with only eight in the control group, giving a relative risk after adjustment for sex of 5.4 (95 per cent CI 2.7–10.8). The average total thyroid dose is estimated to have been about 0.1 Gy in this group, and a risk coefficient of 7.7 per 10^4 person-year Gy (95 per cent confidence interval 3.0–17.2) has been calculated for these data by Shore et al. (1986). This estimate is somewhat higher than the estimates obtained from the group irradiated for an enlarged thymus. The difference might be due to chance, but the risk appeared to be concentrated among Israelis of North African origin, and another possible explanation that has been suggested is that there might be some special susceptibility to tumour induction among this group.

Irradiation of the breast

Five studies of women who received breast irradiation during the course of treatment for benign conditions, together with other studies such as those of atomic bomb survivors, have helped to show that the female breast is one of the most radiosensitive organs in the body. Two of these studies are of women who received X-ray therapy for conditions of the breast itself. Shore et al. (1977) studied about 600 women who received treatment in the 1940s and early 1950s in Rochester, New York, for acute postpartum mastitis (infectious breast disease during childbirth and lactation), together with a control group of 1300, comprising partly women with mastitis who received no X-ray treatment, and partly sisters of both groups of mastitis patients. The mean breast dose in the irradiated group was about 2.5 Gy. During a follow-up period of up to 37 years, 57 breast cancer cases have been observed in the irradiated group compared with 61 in the larger control group, giving a relative risk of about 2-fold (Shore 1986). Baral et al. (1977) have studied about 1020 women in Stockholm who received radiotherapy for fibroadenomatosis and other benign breast conditions. The median dose was about 7 Gy. A total of 115 breast cancers were observed in a follow-up period averaging 32 years, giving a relative risk of 4 compared with general population rates. Although there is no control group for this series, many of the women had received treatment to one breast only and in these women no increase in cancer was seen in the contralateral, untreated breast. Two further studies involved women who received multiple fluoroscopies (X-ray imaging) during pneumotherapy (induced lung collapse) for tuberculosis; women with tuberculosis who received no fluoroscopy served as controls. One group, from Massachusetts, included about 1050 women who received an average breast dose of about 1.5 Gy. Follow-up times averaged 27 years and 41 breast cancers were observed in the irradiated group, giving a relative risk of 1.8 compared with the controls (Boice and Monson 1977). The second study includes about 11 300 women with tuberculosis treated in sanatoria throughout Canada, and about 12 000 unexposed tuberculous controls (Howe 1984). The women were treated during the 1930s and 1940s, and breast cancer deaths were recorded for the period 1950–1977. Among the unexposed women, mortality was very similar to that in the Canadian population, but among those who received fluoroscopies there was an increase in breast cancer mortality of about 60 per cent (based on 172 deaths). Finally in the study of persons exposed in infancy for thymic enlargement and described above, the median breast dose was about 0.1 Gy and a relative risk of 5 (based on nine cases) was seen in the irradiated group compared with their untreated siblings (Hildreth et al. 1985).

There is good agreement between the results of these studies of radiation-induced breast cancer. The minimum latent period appears to be about 10 years, except for those irradiated in infancy for thymic enlargement, where the minimum latent period is about 30–35 years. This confirms the finding from the studies of Japanese atomic bomb survivors, that radiation does not cause the risk of breast cancer to increase before the age at which there is an appreciable breast cancer rate in the general female population at large. It is not yet clear whether the increased risk of breast cancer will last throughout life, but in spite of follow-up periods of 30–40 years, all the above studies still show an increase in risk at the longest follow-up times. Four of the five studies listed above include a wide range of breast doses, and it seems clear that the risk of breast cancer tends to increase with increasing breast dose, although it is not yet clear whether the relationship is linear or whether it includes some curvature.

The total dose of radiation is often delivered in fractional doses, given on several occasions separated in time in order to reduce acute side-effects such as skin burns. Shore (1986) considered the effect of this dose fractionation in the studies of breast cancer. In the mastitis study and the Swedish radiotherapy series, the radiation was received in a few fractions, with typical breast doses per fraction of about 1 Gy, and the risks of developing a breast cancer in these studies were about 8.3 and 6.8 per 10^4 person-year Gy, respectively. These are quite consistent with the risks observed in the studies of multiple fluoroscopies, where the dose was typically received in 100 or so fractions with typical breast doses per fraction of 0.002–0.2 Gy: the risk of developing a breast cancer was 6.2 per 10^4 person-year Gy in Massachusetts, and of dying from breast cancer was 8.4 and 2–4 per 10^4 person-year in Nova Scotia and the rest of Canada respectively. Thus there does not seem to be any appreciable difference in breast cancer risk associated with dose fractionation.

Age at irradiation appears to affect the magnitude of the subsequent risk, with higher risks being observed among those irradiated at younger ages, especially at ages less than 20. The effect of breast irradiation on women aged 40 and above is of special interest in trying to formulate policies for mammographic screening for breast cancer, in which doses of the order of 0.01 Gy would be involved. Unfortunately, the numbers of women irradiated in these older age groups are very small in all the studies.

The Massachusetts fluoroscopy study included only 58 irradiated patients in this age group and the mastitis study only 14 (Boice *et al.* 1979). The large Canadian study does not show any increase in risk for those aged over 40 at irradiation (one death observed compared with 1.16 expected based on population rates; Shore *et al.* 1985). However, the Swedish study indicates that some risk may be incurred after irradiation above the age of 40. Clearly further studies of women irradiated in this age group are needed.

Patients treated with ^{131}iodine

Several studies have been carried out of patients treated with ^{131}iodine for hyperthyroidism (Dobyns *et al.* 1974; Hoffman *et al.* 1982*a*, *b*; Hoffman 1984; Holm 1984). Holm (1984) included just over 4500 patients treated between 1951 and 1975 at the Radiumhemmet hospital in Stockholm, Sweden. The patients received radiation doses of between 60 and 100 Gy to the thyroid, and were followed up for a mean of 9.5 years. Cancer incidence data were obtained by searching the Swedish Cancer Registry. Overall, the number of tumours observed (398) was almost exactly equal to the number expected (395.6). When the data were subdivided into 13 different sites of cancer, including leukaemia, thyroid cancer and cancers of sites exposed to relatively high concentrations of ^{131}iodine during treatment, there were no significant excesses. When the data were further subdivided by sex there was a significant increase of CNS tumours among women (17 observed against 9.0 expected); this could well have been a chance finding. There was no increased risk of cancer among those given larger doses of ^{131}iodine compared with those given smaller doses.

One difficulty with interpreting the results of the Swedish study is the short follow-up time available at present, and the same applies to a large study of 19 200 patients treated in the US, followed up for an average of 8 years, and in whom no increase in leukaemia or thyroid cancer has been seen (Dobyns *et al.* 1974; Saenger 1968). The follow-up of about 1000 of these patients has been extended to an average 15 years. When their mortality was compared with that in a similar sized control group of women treated surgically no increase in cancer risk was observed (Hoffman *et al.* 1982*a*). For cancer incidence an initial analysis reported an elevated risk of cancer (relative risk 1.8, 95 per cent CI 1.1–3.2) among organs that concentrate ^{131}iodine (salivary glands, digestive tract, kidney, and bladder) (Hoffman *et al.* 1982*b*), but in a later report in which organs were ranked according to the estimated doses received, it was concluded that no clear patterns emerged (Hoffman 1984). Further follow-up of these and other populations exposed to ^{131}iodine is clearly required.

Table 3.5 Numbers of patients referred to a radiotherapy centre in Oxford in 1987, by the condition for which they were treated. (Figures are those of Dr. A. Laing, personal communication)

Condition for which treatment given	Number of patients referred
Malignant conditions	2151
Benign conditions	166
Thyrotoxicosis	113
Bone and joint problems (incl. two ankylosing spondylitis)	21
Benign tumours (incl. ten pituitary adenomas)	17
Skin disorders (incl. eight keloids)	12
Benign gynaecological disease (i.e. artificial menopause)	2
Others	1
Referred but no treatment given	28
Total patients referred	2345

Conclusion

At the present time medical radiology is still by far the most important source of exposure to radiation of man-made origin in most populations. For example, in the UK, the average annual effective dose equivalent is estimated to be about 250 μSv, comprising nearly 90 per cent of the total from artificial sources, and just over 10 per cent of the total from all sources (National Radiological Protection Board 1987). Crude estimates have suggested that recent medical uses of radiation might be responsible for about 0.5 per cent of all cancers (Doll and Peto 1981). In the UK, it is estimated that nearly 90 per cent of the population dose from medical radiology is received from diagnostic radiology (Hughes and Roberts 1984) in which the dose to each individual patient is extremely low. Radiotherapy is now largely confined to the treatment of malignant disease, and although both ankylosing spondylitis and benign gynaecological disorders are still sometimes treated by radiotherapy, such treatment is normally used only as a last resort, for example in spondylitis patients for whom other methods of treatment have not provided relief, or in women who are unsuitable for surgery and cannot have a hysterectomy. The situation is illustrated in Table 3.5 by figures from a radiotherapy department in Oxford showing that, in 1987 out of 2345 patients referred for treatment, only 166 (7 per cent) were treated for benign conditions, and of these nearly 70 per cent were given ^{131}iodine for thyrotoxicosis. Thyrotoxicosis is the only benign condition in the UK for which treatment with radiation remains in widespread use (Saenger et al. 1968). A nation-wide survey has estimated that in 1982 it ranked tenth among procedures involving administration of radioactive substances in the UK, involving about 7600 administrations, or approximately 2 per cent of all nuclear medicine administrations (Wall et al. 1985).

References

Baral, E., Larsson, L.-E., and Mattsson, B. (1977). Breast cancer following irradiation of the breast. *Cancer*, **40**, 2905–10.

Boice, J. D., Jr and Monson, R. (1977). Breast cancer in women after repeated fluoroscopic examinations of the chest. *Journal of the National Cancer Institute*, **59**, 823–32.

Boice, J. D., Jr, Land, C. E., Shore, R. E., Norman, J. E., and Tokunaga, M. (1979). Risk of breast cancer following low-dose radiation exposure. *Radiology*, **131**, 589–97.

Chmelevsky, D., Dellerer, A. M., Spiess, H., and Mays, C. W. (1986). A proportional hazards analysis of bone sarcoma rates in German ^{224}radium patients. In *The radiobiology of radium and Thorotrast*, (ed. W. Gössner, G. B. Gerber, U. Hagen, and A. Luz), pp. 32–7. Urban and Schwarzenberg, München.

Court-Brown, W. M. and Doll, R. (1957). *Leukaemia and aplastic anaemia in patients irradiated for ankylosing spondylitis*. HMSO, London.

Court-Brown, W. M. and Doll, R. (1965). Mortality from cancer and other causes after radiotherapy for ankylosing spondylitis. *British Medical Journal*, **2**, 1327–32.

Darby, S. C. and Doll, R. (1988). Trends in long term mortality in ankylosing spondylitics treated with a single course of X-rays. In *Health effects of low dose ionising radiation*, pp. 51–6. BNES, London.

Darby, S. C., Nakashima, E., and Kato, H. (1985). A parallel analysis of cancer mortality among atomic bomb survivors and patients with ankylosing spondylitis given X-ray therapy. *Journal of the National Cancer Institute*, **72**, 1–21.

Darby, S. C., Doll, R., Gill, S. K., and Smith, P. G. (1987). Long term mortality after a single treatment course with X-rays in patients treated for ankylosing spondylitis. *British Journal of Cancer*, **55**, 179–90.

Dobyns, B. M., Sheline, G. E., Workman, J. B., Tompkins, E. A., McConahey, W. M., and Becker, D. V. (1974). Malignant and benign neoplasms of the thyroid in patients treated for hyperthyroidism: a report of the cooperative thyrotoxicosis therapy follow-up study. *Journal of Clinical Endocrinology and Metabolism*, **38**, 976–98.

Doll, R. and Peto, R. (1981). *The causes of cancer*. Oxford University Press.

Doll, R. and Smith, P. G. (1968). The long term effects of X-irradiation in patients treated for metropathia haemorrhagica. *British Journal of Radiology*, **41**, 362–8.

Faber, M. (1983). Current (1981) status of the Danish Thorotrast study. *Health Physics*, **44** (Suppl. 1), 259–60.

Faber, M. (1986). Observations on the Danish Thorotrast patients. *Strahlentherapie*, **80** (Suppl.), 140–2.

Falk, H., Telles, N. C., Ishak, K. G., Thomas, L. B., and Popper, H. (1979). Epidemiology of Thorotrast-induced hepatic angiosarcoma in the United States. *Environmental Research*, **18**, 65–73.

Hickling, P. and Wright, G. (1983). Seronegative arthritis. In *Oxford textbook of medicine*, (ed. D. J. Weatherall, J. G. G. Ledingham, and D. A. Warrell), pp. 16.13–16.22. Oxford University Press.

Hildreth, N. G., Shore, R. E., Hempelmann, L. H., and Rosenstein, M. (1985). Risk of extrathyroid tumours following radiation treatment in infancy for thymic enlargement. *Radiation Research*, **102**, 378–91.

Hoffman, D. (1984). Effects of I-131 therapy in the United States. In *Radiation carcinogenesis: epidemiology and biological significance*, (ed. J. D. Boice, Jr and J. F. Fraumeni, Jr), pp. 273–80. Raven Press, New York.

Hoffman, D. A., McConahey, W. M., Diamond, E. L., and Kurland, L. T. (1982a). Mortality in women treated for hyperthyroidism. *American Journal of Epidemiology*, **115**, 243–54.

Hoffman, D. A., McConahey, W. M., Fraumeni, J. F., Jr, and Kurland, L. T. (1982b). Cancer incidence following treatment of hyperthyroidism. *International Journal of Epidemiology*, **11**, 218–24.

Holm, L.-E. (1984). Malignant disease following iodine-131 therapy in Sweden. In *Radiation carcinogenesis: epidemiology and biological significance*, (ed. J. D. Boice, Jr and J. F. Fraumeni, Jr), pp. 263–271. Raven Press, New York.

Horta, J. S., Horta, M. E., da Motta, L. C., and Tavares, M. H. (1978). Malignancies in Portuguese Thorotrast patients. *Health Physics*, **35**, 137–52.

Howe, G. R. (1984). Epidemiology of radiogenic breast cancer. In *Radiation carcinogenesis: epidemiology and biological significance*, (ed. J. D. Boice, Jr and J. F. Fraumeni, Jr), pp. 119–29. Raven Press, New York.

Hughes, J. S. and Roberts, G. C. (1984). *The radiation exposure of the UK population—1984 review*, NRPB-R173. HMSO, London.

Kaul, A. and Noffz, W. (1978). Tissue dose in Thorotrast patients. *Health Physics*, **35**, 113–21.

Lewis, C. A., Smith, P. G., Stratton, I. M., Darby, S. C., and Doll, R. (1988). Estimated radiation doses to different organs among patients treated for ankylosing spondylitis with a single course of X-rays. *British Journal of Radiology*, **61**, 212–20.

Mays, C. W. (1973). Cancer induction in man from internal radioactivity. *Health Physics*, **25**, 585–92.

Mays, C. W. and Spiess, H. (1984). Bone sarcomas in patients given radium-224. In *Radiation carcinogenesis: epidemiology and biological significance*, (ed. J. D. Boice, Jr and J. F. Fraumeni, Jr), pp. 241–52. Raven Press, New York.

Mays, C. W., Spiess, H., Chmelevsky, D., and Kellerer, A. (1986). Bone sarcoma cumulative tumour rates in patients injected with ^{224}Ra. In *The radiobiology of radium and Thorotrast*, (ed. W. Gössner, G. B. Gerber, U. Hagen, and A. Luz), pp. 27–31. Urban and Schwarzenberg, München.

Mori, T., Kato, Y., Kumatori, T., Maruyama, T., and Hatakeyama, S. (1983). Epidemiological follow-up study of Japanese Thorotrast cases—1980. *Health Physics*, **44** (Suppl. 1), 261–72.

Mori, T., et al. (1986). Present status of medical study of Thorotrast-administered patients in Japan. *Strahlentherapie*, **80** (Suppl.), 123–34.

National Radiological Protection Board (1987). *Living with radiation*. HMSO, London.

National Research Council Committee on the Biological Effects of Ionizing Radiations (1980). *The effects on populations of exposure to low levels of ionizing radiation*. National Academy Press, Washington.

National Research Council Committee on the Biological Effects of Ionizing Radiations (1988). *Health risks of radon and other internally deposited alpha-emitters*. National Academy Press, Washington.

Palmer, J. P. and Spratt, D. W. (1956). Pelvic carcinoma following irradiation for benign gynecological diseases. *American Journal of Obstetrics and Gynecology*, **72**, 497–505.

Pochin, E. (1983). *Nuclear radiation: risks and benefits*. Clarendon Press, Oxford.

Preston, D. L. and Pierce, D. A. (1988). The effect of changes in dosimetry on cancer mortality risk estimates in the atomic bomb survivors. *Radiation Research*, **114**, 437–66.

Radford, E. P., Doll, R., and Smith, P. G. (1977). Mortality among patients with ankylosing spondylitis not given X-ray therapy. *New England Journal of Medicine*, **297**, 572–6.

Ron, E. and Modan, B. (1980). Benign and malignant thyroid neoplasms after childhood irradiation for tinea capitis. *Journal of the National Cancer Institute*, **65**, 7–11.

Ron, E. and Modan, B. (1984). Thyroid and other neoplasms following childhood scalp irradiation. In *Radiation carcinogenesis: epidemiology and biological significance*, (ed. J. D. Boice, Jr and J. F. Fraumeni, Jr), pp. 139–51. Raven Press, New York.

Saenger, E. L., Thoma, G. E., and Tompkins, E. A. (1968). Incidence of leukaemia following treatment of hyperthyroidism: preliminary report of the cooperative thyrotoxicosis therapy follow-up study. *Journal of the American Medical Association*, **205**, 147–54.

Shore, R. E. (1986). Carcinogenic effects of radiation on the human breast. In *Radiation carcinogenesis*, (ed. A. C. Upton, R. E. Albert, F. J. Burns, and R. E. Shore), pp. 280–91. Elsevier Press, New York.

Shore, R. E., et al. (1977). Breast neoplasms in women treated with X-rays for acute postpartum mastitis. *Journal of the National Cancer Institute*, **59**, 813–22.

Shore, R. E., Albert, R. E., Reed, M., Harley, N., and Pasternack, B. S. (1984). Skin cancer incidence among children irradiated for ringworm of the scalp. *Radiation Research*, **100**, 192–204.

Shore, R. E., Woodward, E., Hildreth, N., Dvoretsky, P., Hempelman, L., and Pasternack, B. (1985). Thyroid tumours following thymus irradiation. *Journal of the National Cancer Institute*, **74**, 1177–84.

Shore, R. E., Hempelman, L. H., and Woodward, E. D. (1986). Carcinogenic effects of radiation on the human thyroid gland. In *Radiation carcinogenesis*, (ed. A. C. Upton, R. E. Albert, F. J. Burns, and R. E. Shore), pp. 293–309. Elsevier Press, New York.

Simon, N., Brucer, M., and Hayes, R. (1960). Radiation and leukaemia in cancer of the cervix. *Radiology*, **74**, 905–13.

Smith, P. G. and Doll, R. (1976). Late effects of X-irradiation in patients treated for metropathia haemorrhagica. *British Journal of Radiology*, **49**, 224–32.

Smith, P. G. and Doll, R. (1982). Mortality among patients with ankylosing spondylitis after a single treatment course with X-rays. *British Medical Journal*, **284**, 449–60.

Smith, P. G., Doll, R., and Radford, E. P. (1977). Cancer mortality among patients with ankylosing spondylitis not given X-ray therapy. *British Journal of Radiology*, **50**, 728–34.

van Kaick, G., *et al.* (1984). Results of the German Thorotrast study. In *Radiation carcinogenesis: epidemiology and biological significance*, (ed. J. D. Boice, Jr and J. F. Fraumeni, Jr), pp. 253–62. Raven Press, New York.

van Kaick, G., *et al.* (1986). Report on the German Thorotrast study. *Strahlentherapie*, **80** (Suppl.), 114–8.

Vaughan, J. (1986). Carcinogenic effects of radiation on the human skeleton and supporting tissues. In *Radiation carcinogenesis*, (ed. A. C. Upton, R. E. Albert, F. J. Burns, and R. E. Shore), pp. 311–44. Elsevier Press, New York.

Wagoner, J. K. (1969). *Leukaemia and other malignancies following radiation therapy for benign gynecological disorders.* Doctoral dissertation, Harvard School of Public Health, Boston.

Wagoner, J. K. (1984). Leukemia and other malignancies following radiation therapy for gynecological disorders. In *Radiation carcinogenesis: epidemiology and biological significance*, (ed. J. D. Boice, Jr and J. F. Fraumeni, Jr), pp. 153–9. Raven Press, New York.

Wall, B. F., Hillier, M. C., Kendall, G. M., and Shields, R. A. (1985). Nuclear medicine activity in the United Kingdom. *British Journal of Radiology*, **58**, 125–30.

Wick, R. R. and Gössner, W. (1983). Follow-up study of late effects in ^{224}Ra treated ankylosing spondylitis patients. *Health Physics*, **44** (Suppl. 1), 187–95.

Wick, R. R., Chmelevsky, D., and Gössner, W. (1986). ^{224}Ra: risk to bone and haematopoietic tissue in ankylosing spondylitis patients. In *The radiobiology of radium and Thorotrast*, (ed. W. Gössner, G. B. Gerber, U. Hagen, and A. Luz), pp. 38–44. Urban and Schwarzenberg, München.

4
Cytotoxic chemotherapy for cancer

JOHN M. KALDOR and CHRISTINE LASSET

Introduction

Cytotoxic chemotherapy has become a standard part of the treatment for a number of forms of cancer over the past few decades. It has radically reduced mortality from Hodgkin's disease, acute lymphocytic leukaemia in children, Wilms' tumour, retinoblastoma, testicular teratoma, choriocarcinoma, and several types of non-Hodgkin's lymphoma. Given as adjuvant therapy following surgery for cancers of the breast (Consensus Conference 1985) and ovary, chemotherapy appears to improve survival. For many other types of malignancy, cytotoxic chemotherapy has not been shown to extend survival time, although it may reduce tumour size in the short-term.

The several dozen drugs which have been used for cancer chemotherapy can be classified according to their mechanism of cytotoxicity. The biggest group, which was also the first to be developed, is made up of the alkylating agents. These compounds bind covalently to cellular DNA, either directly or following metabolic activation, and their cytotoxicity is probably a result of inter- and intra-strand crosslinks (Hemminki and Ludlum 1984). It is now well established that most, if not all, of the alkylating chemotherapeutic agents are carcinogenic. The International Agency for Research on Cancer has evaluated 17 alkylating agents for carcinogenicity in humans and experimental animals (IARC 1987a, 1990), and the results are summarized in Table 4.1.

Table 4.1 also includes cytotoxic drugs which act by mechanisms other than DNA alkylation, including interference with mitosis, inhibition of cellular metabolism, and intercalation in the DNA molecule. None of these agents has been shown to be carcinogenic in humans, although, like the alkylating agents, several of them are active in a variety of short-term tests for mutagenicity, clastogenicity or related properties (IARC 1987b), and two (actinomycin D and adriamycin) have carcinogenic activity in animal bioassays. Evaluation of their carcinogenic effect in humans is complicated by the fact that they are usually given in combination, with or without alkylating agents. This difficulty has now also come to apply to the study of the alkylating agents, and most human data implicating specific

Table 4.1 Antineoplastic drugs evaluated in IARC Monographs

Mechanism of cytotoxic action	Drug name	Evidence for carcinogenicity[1]	
		Humans	Animals
DNA alkylation	Busulfan (1,4-butanediol dimethane-sulphonate)	S	L
	Chlorambucil	S	S
	Chlornaphazine (N,N-bis(2-chloro-ethyl)-2 naphthylamine	S	L
	Cyclophosphamide	S	S
	Melphalan	S	S
	Semustine (methyl-CCNU)	S	L
	Treosulphan	S	ND
	BCNU (bis-chloroethylnitrosourea)	I	S
	CCNU (1-(2-chloroethyl)-3-cyclohexyl-1-nitrosourea)	I	S
	Cisplatin	I	S
	Dacarbazine	I	S
	Mitomycin C	I	S
	Nitrogen mustard	I	S
	Procarbazine	I	S
	Thio-TEPA (tris(1-aziridinyl)-phosphine sulphide)	S	S
	Uracil mustard	I	S
	Isophosphamide	ND	L
Anti-metabolite	5-Fluorouracil	I	I
	6-Mercaptopurine	I	I
	Methotrexate	I	I
Mitosis inhibition	Vinblastine	I	I
	Vincristine	I	I
DNA intercalation and strand breakage	Actinomycin D	I	L
	Adriamycin	I	S
DNA strand breakage	Bleomycin	I	I

[1] S = Sufficient evidence for carcinogenicity; L = limited evidence for carcinogenicity; I = inadequate evidence for carcinogenicity; ND = no adequate data for an evaluation. For further details, see IARC (1987a, 1990).

agents as carcinogens come from earlier periods and relate to drugs which were used singly (Kaldor et al. 1988).

Several aspects of cancer following cytotoxic chemotherapy will be considered in this review, including the types of tumour produced, the relative potency of different drugs and combinations, the interaction with radiotherapy and other factors, and the temporal aspects of second cancer risk. Animal bioassays for cytotoxic drugs (reviewed by Berger 1986) and short-term tests (IARC 1987*b*; Fishbein 1987) are not considered further. Other subjects beyond the scope of this chapter are the diagnosis, characterization, and treatment of cancers induced by chemotherapy. Immunosuppressive agents are considered in detail in Chapter 5 (Dorreen and Hancock, this volume).

Sources of information

In describing various aspects of chemotherapy-induced second cancer, we have drawn upon published studies which satisfied certain criteria. First, case reports of the occurrence of a second cancer in one or more treated patients have generally been excluded if no corresponding information on the population at risk was provided: such reports clearly play a crucial role in clinical research, but cannot be used to establish causality or to study quantitative issues. Second, we required that the study reported second cancer risk in a clearly identifiable group of patients who had been treated with chemotherapy, and that comparison had been made with an appropriate reference group of patients, either not treated with chemotherapy, or treated with other kinds of chemotherapy. Clearly, the ideal source of information would be a clinical trial in which comparisons were made of second cancer risk between the arms of the trial. However, most published studies have been historical comparisons or, more recently, have adopted the case-control design to estimate the relative risk of second cancer due to different types of treatment. A number of studies have used, as a comparison group, the general population of the same sex and age and in the same calendar period as the cancer patients under investigation for second cancer risk. Kaldor et al. (1988) discuss the relative merits of these various options for studying second cancer risk.

Leukaemia

Leukaemia is by far the most frequently reported malignancy following chemotherapy and the most clearly established as causally associated. Although it was originally thought to be a late complication of advanced-stage Hodgkin's disease apparently cured by multiple-agent chemotherapy, subsequent studies have established the role of chemotherapy itself,

Table 4.2 Risk of acute non-lymphocytic leukaemia following Hodgkin's disease: results summarized from three studies[1]

Category of treatment	Number of cases of ANLL	Relative risk Compared to general population	Compared to radiotherapy only
Radiotherapy only	4	4.9	1.0[2]
Chemotherapy only	11	120	24
Radiotherapy and chemotherapy	42	130	26

[1] Boivin et al. 1984; Tucker et al. 1987a; Colman et al. 1988.
[2] By definition.

by comparing leukaemia risk among groups of patients treated with chemotherapy, radiotherapy and both types of therapy (Arseneau et al. 1972). Table 4.2 summarizes the estimated relative risk of leukaemia associated with these three categories of treatment from three large studies of Hodgkin's disease survivors. While the leukaemia risk following chemotherapy for Hodgkin's disease is very large, both relative to the general population and relative to patients treated only with radiotherapy, it should be kept in mind that before the advent of multiple agent chemotherapy, advanced-stage Hodgkin's disease was almost uniformly fatal: there is thus no way to know what the risk of leukaemia would be in the absence of chemotherapy. A notable feature of Table 4.2 is the magnitude of the relative risk for radiotherapy alone, compared to the general population, suggesting that radiation is more leukaemogenic in Hodgkin's disease patients than in survivors of cervical and other cancers (Boice et al. 1985).

For several other types of cancer, an appreciable fraction of patients have been treated by surgery alone, and the leukaemia risk among these patients appears to be no higher than would be expected on the basis of general population rates. Table 4.3 indicates the risk relative to the general population of leukaemia among survivors of ovarian cancer, breast cancer, small-cell carcinoma of the lung, and gastrointestinal cancer who were treated with chemotherapy. The table also gives relative risk estimates for leukaemia following chemotherapy for non-Hodgkin's lymphoma and multiple myeloma. Survivors of childhood leukaemia treated with alkylating agents (Tucker et al. 1987a) are a further group who have been observed to be at a substantially increased risk of leukaemia. Several of the relative risks in Table 4.3 are of a magnitude comparable to those for Hodgkin's disease patients treated with chemotherapy (Table 4.2). Even discounting substantially for more diligent case-finding in cancer patients as compared to the general population, and for the possibility that tumour types such as non-Hodgkin's lymphoma and multiple myeloma may predispose to

Table 4.3 Acute non-lymphocytic leukaemia (ANLL) following chemotherapy for various types of cancer

First cancer	Observed cases of ANLL	Relative risk compared to general population	Type of chemotherapy	References
Ovary	17	100	Melphalan and cyclophosphamide	Greene et al. 1986
Breast	7	170	Treosulphan	Pedersen-Bjergaard et al. 1980
	26	24	Melphalan	Fisher et al. 1985
Lung[1] (small cell carcinoma)	8	130	Multiple agents	Pedersen-Bjergaard et al. 1985a Chak et al. 1984 Johnson et al. 1986
Non-Hodgkin's lymphoma[2]	13	65	Multiple agents	Pedersen-Bjergaard et al. 1985b Greene et al. 1983
Gastrointestinal cancer	7	9.9	Semustine and 5-fluorouracil	Boice et al. 1983
Multiple myeloma	14	210	Multiple agents	Bergsagel et al. 1979

[1] Results combined from three studies.
[2] Results combined from two studies.

leukaemia, it is difficult to attribute risk increases of this size to anything but chemotherapy.

Leukaemia subtypes

Most of the leukaemias observed following chemotherapy are of the acute myeloid subtype, but it appears that other acute non-lymphocytic types are also produced. The classification 'acute non-lymphocytic leukaemia' (ANLL) has become the accepted designation for the category of leukaemias related to alkylating agent chemotherapy. Table 4.4 shows the numbers of leukaemias of different types observed within 10 years of Hodgkin's disease and ovarian cancer in two large international case-control studies (Kaldor et al. 1990a,b).

Table 4.4 Number of cases of leukaemia by subtype and time since first cancer following Hodgkin's disease and ovarian cancer[1]

Leukaemia subtype	Time since diagnosis of first cancer (years)			
	Hodgkin's disease		Ovarian cancer	
	1–5	6 +	1–5	6 +
Acute				
Myeloid	51	38	30	18
Monocytic	9	4	1	2
Lymphocytic	2	2	4	2
Other and unspecified	7	10	12	3
Erythroleukaemia	6	1	5	2
Other non-lymphocytic	3	5	7	5
Chronic lymphocytic	1	1	3	4
Leukaemia, unspecified	8	4	2	3
Total	87	65	64	39

[1] Kaldor et al. 1990a,b.

Generally, there have been few cases of leukaemia of types other than ANLL observed following chemotherapy, and probably not enough to exclude the possibility that their incidence is increased by chemotherapy. There is, however, a clear increase in the incidence of the myelodysplastic syndrome (Bennett et al. 1981), a condition which sometimes leads to leukaemia, and indeed has been referred to as preleukaemia, but which is itself frequently fatal. A number of recent studies of chemotherapy-induced malignancies have grouped the cases of myelodysplastic syndrome together with the cases of leukaemia.

Temporal factors

The leukaemias induced by chemotherapy can appear within 1 or 2 years of treatment, but the peak in excess incidence seems to come 5 to 7 years after the first course of chemotherapy. Very few studies have formally estimated the excess or relative incidence by time since treatment began, although several give the cumulative leukaemia risk as a function of time (Kaplan–Meier curve). Figure 4.1 shows the absolute incidence of leukaemia as a function of time since the beginning of treatment for Hodgkin's disease (Blayney *et al.* 1987), and in Fig. 4.2, the time pattern of relative risk of leukaemia due to chemotherapy, compared to patients who never received chemotherapy, is estimated from recent case-control studies of leukaemia following ovarian cancer (Kaldor *et al.* 1990*a*) and Hodgkin's disease (Kaldor *et al.* 1990*b*). Although the relative risk of leukaemia appears to remain elevated for at least 10 years since the start of chemotherapy, the risk clearly diminishes after the initial increase, in a manner reminiscent of the excess leukaemia risk observed following exposure to ionizing radiation (UNSCEAR 1986).

Figure 4.2 also shows the relative risk by time since the last chemotherapy, for survivors of Hodgkin's disease and ovarian cancer. There is a suggestion that the excess risk disappears within a decade of cessation of treatment, but the number of patients who have been observed for this length of time is relatively small, and does not permit a precise estimation of risk.

In a study of patients treated with chemotherapy for multiple myeloma,

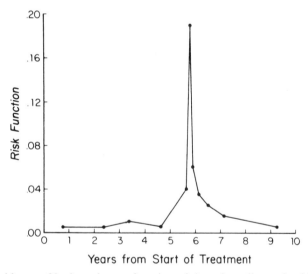

Fig. 4.1 Incidence of leukaemia as a function of time since diagnosis of Hodgkin's disease (from Blayney *et al.* 1987, reprinted, with permission, from the *New England Journal of Medicine*, **316**, 710–14, 1987).

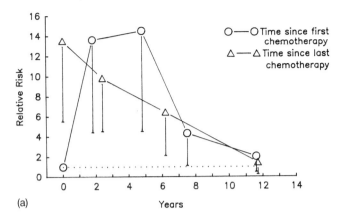

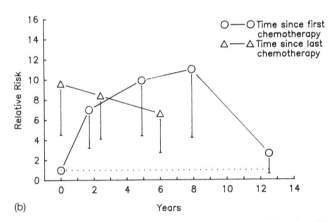

Fig. 4.2 Relative risk of acute non-lymphocytic leukaemia following (a) ovarian cancer and (b) Hodgkin's disease, as a function of time since first chemotherapy and time since last chemotherapy (from Kaldor *et al.* 1990a,b).

Cuzick *et al.* (1987) found that the amount of melphalan given in the most recent 3-year period was the most important determinant of the risk of developing myelodysplasia or acute myeloid leukaemia. After accounting for melphalan in this period, earlier chemotherapy did not appear to be of importance in predicting which patients would develop leukaemia, although the small number of cases in the study (12) suggests that this finding should be interpreted cautiously.

Leukaemogenic potency

Several studies have permitted a comparison of leukaemogenic potency between two or more drugs. Potency can be simply defined as an increase in leukaemia risk either per unit total dose, or per unit of equal therapeutic effect. However, its estimation in either case requires a considerable number of assumptions, including linearity of the leukaemogenic effect and its independence of age, sex or other factors specific to each patient. Comparisons of potency estimates between studies generally involve the further assumption that the type of first cancer has no effect on leukaemia risk. Kaldor et al. (1988) have estimated leukaemogenic potency in humans from published studies of cytotoxic drugs which are used or have been used in cancer chemotherapy. With the above caveats in mind concerning the estimation of potency, chlorambucil, semustine, and melphalan appeared to be the most potent leukaemogens on a per gram basis, and cyclophosphamide and treosulphan the weakest. Two studies provide information on both cyclophosphamide and melphalan used in the same clinical trial, and they demonstrate that melphalan is the more potent leukaemogen (Green et al. 1986; Cuzick et al. 1987). No leukaemias have been reported so far among several hundred women treated with methotrexate as a single agent for choriocarcinoma (Rustin et al. 1983; Trapido, pers. commn). A recent case-control study of ovarian cancer patients (Kaldor et al. 1990a) has allowed comparison of the leukaemogenicity of five alkylating agents (see Table 4.5).

The substantial variation in potency among the different agents draws attention to a potential difficulty in the use of alkylator scores. These scores have been used in several studies to summarize the amount of chemotherapy received by each patient (Greene et al. 1986; Tucker et al. 1987a,b). A patient's alkylator score is calculated by replacing the dose of each drug received by its percentile in the distribution of doses over all patients who received that drug, and summing these percentiles for each drug received. Such scores are not strictly appropriate when the drugs under consideration are of different leukaemogenic potency. Nevertheless, the alkylator score can still provide an approximate index of the amount of drug received. Table 4.6 gives the estimate of leukaemogenic relative risk for increasing categories of dose for some selected drugs.

For combination chemotherapy, which is given in several cycles separated by short periods without treatment, it would be meaningless to make potency comparisons per unit weight of the drugs in the combination, so a simpler approach based on the number of cycles received has been adopted. In a case-control study of leukaemia following Hodgkin's disease (Kaldor et al. 1990b), it was found that the risk of leukaemia associated with more than six cycles of chemotherapy containing nitrogen mustard and procarbazine, but with no other alkylating agent, was three times greater than the risk for less than six cycles, and this risk in turn was about five times higher than the risk for patients treated with radiotherapy alone. This category of combination chemotherapy includes MOPP (nitrogen mustard, vincristine, procarbazine, and prednisone) and similar but less

Table 4.5 Relative risk of acute non-lymphocytic leukaemia among ovarian cancer patients who only ever received one alkylating agent[1]

Drug	Low dose[2]			High dose[2]		
	Median dose in controls (g)	Number of cases	Relative risk[3]	Median dose in controls (g)	Number of cases	Relative risk[3]
Chlorambucil	0.17	2	14*	3.2	5	23**
Cyclophosphamide	1.2	4	2.2	22	8	4.1
Melphalan	0.17	9	12*	0.4	17	23**
Thio-TEPA	0.03	4	8.3*	0.6	5	9.7**
Treosulphan	64	1	3.6	260	7	33**

* $p<.05$ (two-sided); ** $p<.01$ (two-sided).
[1] Kaldor et al. 1990a.
[2] Defined by the median dose in all controls who received a given drug: the median dose in controls within the two dose groups so formed is shown in the table.
[3] Relative to patients not treated by chemotherapy.

Table 4.6 Relative risk[1] of leukaemia following treatment with selected chemotherapeutic agents, by dose

Drug	Dose category[2]				Reference
	I	II	III	IV	
Cyclophosphamide	1.5	3.3	7.3		Haas *et al.* 1987
Methyl-CCNU	8.7	10	18	37	Boice *et al.* 1983
Melphalan	11	30	120		Greene *et al.* 1986

[1] Compared to patients who did not receive the drug.
[2] As defined by authors.

widely used combinations, such as MVPP (in which vincristine is replaced by vinblastine). These combinations have been the main form of chemotherapy used for advanced Hodgkin's disease in the past two decades. There is some evidence that recently developed combinations which include adriamycin are less leukaemogenic. Valagussa *et al.* (1986) reported nine leukaemias following MOPP for Hodgkin's disease among 335 patients followed for a median of 9 years, as compared to no cases among 180 patients treated with ABVD (adriamycin, bleomycin, vinblastine, and dacarbazine) and observed for a median of 8.5 years. This finding has been partly responsible for substitution of MOPP by ABVD, or at least alternation of the two combinations, in several more recent treatment protocols. No case of leukaemia was observed among Hodgkin's disease patients treated with ABVD alone in the large case-control study described above (Kaldor *et al.* 1990*b*).

The influence of the first cancer

The influence of the first cancer on the leukaemogenicity of cytotoxic chemotherapy is difficult to investigate, because even if the same drug is used for several types of cancer, it is likely that it is used at different doses, or with different schedules of administration, or in different combinations with other drugs, for each of the various types of cancer. For example, the observation that leukaemia risk following adjuvant melphalan therapy for ovarian cancer was substantially higher (Green *et al.* 1986) than that following similar treatment for breast cancer (Fisher *et al.* 1985)—cumulative 10-year risks of 11.2 per cent and 1.7 per cent, respectively—even though the median cumulative doses of melphalan were approximately the same, can probably be explained in terms of differences in the dose distribution rather than differences due to the first cancers. The leukaemias following ovarian cancer all occurred in patients who had received substantially more than the median dose, while the breast cancer patients, including those who developed leukaemia, all received doses close to the median. Other possible explanations include the longer duration of administration following ovarian cancer, and differences in age between the two groups of patients. Differences in the relative risk of leukaemia following cyclophosphamide for ovarian and breast cancer (Haas *et al.* 1987) may be more easily attributed to differences in the dose of cyclophosphamide.

Interaction of chemotherapy and radiotherapy

An issue of considerable clinical importance is the degree to which chemotherapy and radiotherapy interact in increasing the risk of leukaemia. It is quite clear that chemotherapy with alkylating agents confers a far higher risk than radiotherapy, whether used in the treatment of Hodgkin's disease (see Table 4.2), ovarian cancer (Greene *et al.* 1986), breast cancer (Fisher *et al.* 1985) or other types of cancer. However, several early studies suggested that patients with Hodgkin's disease treated with both chemotherapy and radiotherapy might be at a higher risk of leukaemia than those treated with chemotherapy alone (Coleman *et al.* 1977; Valagussa *et al.* 1980). More recent studies have not supported this suggestion, and have indicated that, after accounting for the total amount of chemotherapy, the dose of radiation does not substantially modify the risk of leukaemia following Hodgkin's disease, although one large study reported a 3-fold increase in risk at intermediate radiation doses (10–20 Gy to the active bone marrow) as compared to lower or higher doses, for approximately the same amount of chemotherapy (Kaldor *et al.* 1990*b*).

Studies of leukaemia following other malignancies have yielded a variety of results. One study of non-Hodgkin's lymphoma patients (Pedersen-Bjergaard *et al.* 1985*b*) found no additional effect of radiotherapy after taking account of chemotherapy, while another (Greene *et al.* 1983) indicated that in each category of chemotherapy, patients who had received high doses of radiotherapy were at a substantially elevated risk of leukaemia when compared to those who had received lower doses of radiation. In patients with ovarian cancer, radiation does not seem to increase the leukaemia risk associated with chemotherapy (Greene *et al.* 1986; Kaldor *et al.* 1990*a*). In summary, the degree to which chemotherapy and radiotherapy interact remains to be clearly elucidated. Statistical analyses in the studies published so far have not adequately accounted for the sequence and timing of the two types of treatment, and the radiation dosimetry could probably be carried out in more detail, for example by estimating dose to a number of body-compartments (Boice *et al.* 1987).

Bladder cancer

The only other form of second cancer which has been clearly and causally linked to cytotoxic chemotherapy has been cancer of the urinary bladder.

In the 1960s, a substantially increased incidence was observed following treatment of polycythaemia vera with chlornaphazine (Thiede and Christensen 1969). This drug, use of which has since been discontinued, is metabolized to yield β-naphthylamine, which, like several other aromatic amines, is recognized as a potent human bladder carcinogen (IARC 1987a).

Cyclophosphamide has also been identified as a human bladder carcinogen (IARC 1987a) on the basis of studies of patients treated with the drug for both malignant and non-malignant diseases. The total number of reported cases is small (in a 1985 review, Seo and colleagues identified only 38 cases), and the association was probably recognized largely because cyclophosphamide was known to cause haemorrhagic cystitis and other acute effects on the bladder. The largest single study to date has included only nine cases of bladder cancer (Pedersen-Bjergaard et al. 1988) and other studies have been based on three or less. The small number of reported cases, which can be ascribed to the relatively small increase in risk and the rather long latency period for bladder cancer compared to acute leukaemia following chemotherapy, limits the precision of the quantitative statements that can be made about cyclophosphamide-induced bladder tumours.

Squamous cell carcinomas of the bladder have been reported following cyclophosphamide treatment (Wall and Clausen 1975), but in the biggest study available, all nine cases of bladder cancer following non-Hodgkin's lymphoma were transitional cell carcinomas (Pedersen-Bjergaard 1988). Of these cases, seven occurred among 471 patients who were known to have received cyclophosphamide, giving a cumulative incidence at 12 years of almost 11 per cent, and a relative risk of 6.8 compared to the general population of Denmark (taking account of age and sex). Similar relative risks have been observed in studies of patients treated for non-malignant disease with cyclophosphamide (Kinlen 1985; Kinlen et al. 1979).

In contrast to leukaemia, the earliest cases of bladder cancer appear to occur at least 6 years after the first treatment with cyclophosphamide, and there is so far no evidence that the excess risk declines or even levels off following the cessation of cyclophosphamide treatment. Bladder cancer also differs from acute leukaemia in that it is several times more frequent in the general population of countries such as Denmark (Muir et al. 1987) where a number of the studies of cyclophosphamide have taken place, and its incidence increases rapidly with age. Assessment of the temporal, and indeed other aspects of cyclophosphamide-induced bladder cancer must therefore take particularly careful account of age. Other factors which may be important are the known bladder carcinogens, such as tobacco smoking and occupational agents. However, in the small studies available, these have not been shown to be related to the risk of bladder cancer following cyclophosphamide therapy, and there is in any case no reason to assume that exposure to these substances might be correlated with the decision to treat a patient with cyclophosphamide.

Many of the reported cases of bladder cancer following cyclophosphamide treatment had also been treated with radiotherapy, but there is so far no information on any interaction between the two types of treatment.

Other cancers

Non-Hodgkin's lymphoma, lung cancer, and several other tumour types have been consistently shown to occur among Hodgkin's disease survivors with a higher frequency than would be expected on the basis of incidence rates of these tumours in the general population (Kaldor et al. 1987; Tucker et al. 1988). Because chemotherapy has played such an important role in improving survival from Hodgkin's disease in recent years, it has also been suspected as being the cause of increased risk of second cancers other than leukaemia and bladder cancer, even though many patients are also treated with X-irradiation, which is known to be carcinogenic. There is so far no clear evidence, however, to implicate chemotherapy for Hodgkin's disease. In a study of over 1500 patients, the relative risk of non-Hodgkin's lymphoma was nearly identical among Hodgkin's disease patients treated with either chemotherapy or radiotherapy, while the numbers of each type of solid tumour were too small to permit a comparison of risks by treatment type (Tucker et al. 1988). Other investigations of Hodgkin's disease patients have encountered similar limitations in the assessment of solid tumour risk, even though several have included follow-up of large numbers of patients (Glicksman et al. 1982; Coltman and Dixon 1982; Pedersen-Bjergaard et al. 1987). A study of lung cancer cases reported after Hodgkin's disease (List et al. 1985) suggested that the risk was nearly six times higher than expected in the general population, and concluded that radiotherapy was a major contributing factor, although the role of chemotherapy could not be excluded.

There have been few investigations of the role of chemotherapy in causing lymphomas and solid tumours following malignancies other than Hodgkin's disease. Studies of tumour incidence following ovarian cancer, testicular cancer and other types of predominantly adult cancer which are frequently treated with chemotherapy have certainly not detected as broad a range of tumours at elevated risk as has been observed following Hodgkin's disease (Curtis et al. 1985; Storm et al. 1985; Kaldor et al. 1987). Childhood cancer survivors appear to be at an increased risk of second malignancy even in the absence of chemotherapy or radiotherapy, but it has not been possible to establish an association with chemotherapy for specific tumour types other than leukaemia (Tucker et al. 1987a) and possibly bone sarcoma (Tucker et al. 1987b).

Nevertheless, as shown in Table 4.7, both Hawkins *et al.* (1987) and de Vathaire *et al.* (1989*a*) have found that after grouping all first and second cancer types together, the relative risk of second cancer, compared to the general population, was higher among children who had been treated with chemotherapy than among those who had not. In a subsequent case-control study within the group of childhood cancer patients, de Vathaire *et al.* (1989*b*) detected an increased risk due to chemotherapy at each of several levels of estimated radiation exposure, a smaller increase in risk associated with cyclophosphamide than with other alkylating agents, and an elevated risk of bone and soft tissue sarcomas in patients who had been treated with actinomycin D. Tucker *et al.* (1987*b*), in a case-control study of bone sarcomas in childhood cancer survivors, found that alkylating agent chemotherapy was associated with about a 5-fold increase in risk. These findings implicating chemotherapy should probably not be viewed as conclusive, since they are based on broad groups of tumour types and, when patients who had not also received radiotherapy are excluded, on rather small numbers of cases.

Table 4.7 Relative risk for all second cancers following childhood cancer, by type of treatment[1]

Reference	Treatment category	
	Radiotherapy	Chemotherapy
Hawkins *et al.* 1987	1.4	1.7
de Vathaire *et al.* 1989*a*	1.2	2.8

[1] Compared to patients who did not receive the treatment.

In summary, it appears that the only causal relationships which have been clearly established are those between alkylating agent chemotherapy and ANLL, and between cyclophosphamide and bladder cancer. It would nevertheless be surprising if the carcinogenicity of cytotoxic agents were quite so specific. A number of cytotoxic agents produce a range of tumours in animal experiments (Berger 1986; IARC 1987*a*), and they are almost all active in several classes of short-term tests (IARC 1987*b*). Since most chemotherapy is administered systemically, all tissues are potentially exposed, although there is considerable variation in exposure levels due to pharmacokinetic and metabolic differences. It may be that the accumulated follow-up time has still been too short, and that a clear excess risk in certain organs will appear only when a substantial number of patients have survived 20 or more years following treatment. Latency periods of this length were required to detect the carcinogenicity of several exposures arising during industrial processes, such as vinyl chloride (Spirtas and

Kaminski 1978). An alternative explanation for the absence, so far, of observations of increased risk at multiple sites, may be that human cells are generally better protected than those of rodents from the alkylating and other genetically damaging effects of cytotoxic drugs. There are certainly differences in the type and quantity of repair enzymes, both between organs and between species (Lindahl 1988), and these may play a role in determining susceptibility to carcinogenesis. Somewhere in between these two extreme explanations lies the possibility that cytotoxic agents are carcinogenic to other human tissues, but to a lesser extent than to the bone marrow and with a longer latent period, and it will simply require several more years of carefully executed studies before these effects can be clearly detected.

Prevention of chemotherapy-induced second cancer

The recognition that intensive alkylating-agent chemotherapy carries a high risk of inducing rapidly fatal acute leukaemia has already resulted in modifications of medical oncological practice. The goal of several recent Hodgkin's disease clinical trials has been to find new combinations of drugs which are as effective against advanced disease as the established ones, but which may be expected to be less leukaemogenic. For example, recent evidence (Carde, pers. commn) suggests that dacarbazine, which is a potent animal carcinogen (IARC 1987a), can be dropped without loss of therapeutic effect from the combination ABVD, which as described earlier, itself seems to be less leukaemogenic than MOPP. On the other hand, procarbazine appears to be an important therapeutic component of MOPP and the hope of reducing the risk of a second tumour now lies in reduction of the number of cycles.

So far, the dramatic success of cisplatinum and related drugs in curing testicular teratoma has not been accompanied by reports of ANLL or other second cancers, even though cisplatinum causes leukaemia and kidney tumours in the rat (Kempf and Ivankovic 1986a,b). Again, it may be that these drugs are still too recently introduced to have been well evaluated, or humans may be in some way less susceptible. On the other hand, there have been recent reports of ANLL following the combination of adriamycin (another important drug which has been shown to be carcinogenic to animals but not yet to humans) and cisplatinum (Pedersen-Bjergaard 1985c; Kaldor et al. 1990a).

The central role of chemotherapy in the treatment of both Hodgkin's disease and non-seminoma testicular cancer is beyond dispute. There are, however, other types of cancer for which its value is less clearly established, and in these cases, long-term risk has arisen as an important argument against its use. For example, adjuvant chemotherapy involving cytotoxic agents appears to improve survival following surgery for stage II premenopausal breast cancer (Consensus Conference 1985), but its use for localized (stage I) tumours may be hard to justify, because survival is already very good, and even a small increase in long-term mortality could outweigh any additional benefit. Ovarian cancer illustrates the problem in another way. Overall survival is poor (5-year relative survival 37 per cent) (NCI 1988), but for the small proportion (23 per cent) of women treated at an early stage, survival is much better (5-year relative survival 84 per cent), and the benefits for such women of prolonged alkylating agent therapy, which is highly leukaemogenic, should now be reevaluated.

There are some promising developments in the synthesis of drugs to counteract the carcinogenic effects of cytotoxic agents (see Fraser, Chapter 8, this volume). Cyclophosphamide-induced cystitis can now be prevented by the simultaneous administration of mesna (Scheef *et al.* 1979), which acts by neutralizing acrolein, a metabolite of cyclophosphamide assumed to be responsible for its toxicity to the urinary tract (Brock *et al.* 1979). Mesna also prevents cyclophosphamide from causing bladder cancer in rats (Habs and Schmähl 1983) and it may have the same effect in humans. Paukowits (1982) describes proteins involved in regulating the division of bone marrow stem cells, which, if given during treatment with alkylating agents, might protect the stem cells, preventing both the acute cytopaenic reaction and the increase in leukaemia risk in the longer-term.

References

Arseneau, J. C., *et al.* (1972). Nonlymphomatous malignant tumors complicating Hodgkin's disease. *New England Journal of Medicine*, **287**, 1119–23.

Bennett, J. M., *et al.* (1981). The morphological classification of acute lymphoblastic leukaemia: concordance among observers and clinical correlations. *British Journal of Haematology*, **47**, 553–61.

Berger, M. R. (1986). Carcinogenicity of alkylating cytostatic drugs in animals. In *Carcinogenicity of alkylating cytostatic drugs*, IARC Scientific Publications No. 78., (ed. D. Schmähl and J. M. Kaldor), pp. 161–76. International Agency for Research on Cancer, Lyon.

Bergsagel, D. E., Bailey, A. J., Langley, G. R., MacDonald, R. N., White, D. F., and Miller, A. B. (1979). The chemotherapy of plasma-cell myeloma and the incidence of acute leukaemia. *New England Journal of Medicine*, **301**, 743–8.

Blayney, D. W., *et al.* (1987). Decreasing risk of leukemia with prolonged follow-up after chemotherapy and radiotherapy for Hodgkin's disease. *New England Journal of Medicine*, **316**, 710–14.

Boice, J. D., *et al.* (1983). Leukemia and preleukemia after adjuvant treatment of gastrointestinal cancer with semustine (methyl-CCNU). *New England Journal of Medicine*, **309**, 1079–84.

Boice, J. D., *et al.* (1985). Second cancers following radiation treatment for cervical cancer. An international collaboration among cancer registries. *Journal of the National Cancer Institute*, **74**, 955–75.

Boice, J. D., et al. (1987). Radiation dose and leukaemia risk in patients treated for cancer of the cervix. *Journal of the National Cancer Institute*, **79**, 1295–311.

Boivin, J. F., Hutchison, G. B., Lyden, M., Goldbold, J., Chorosh, J., and Schottenfeld, D. (1984). Second primary cancers following treatment of Hodgkin's disease. *Journal of the National Cancer Institute*, **72**, 233–41.

Brock, N., Stekar, J., Pohl, J., Niemeyer, U., and Scheffler, G. (1979). Acrolein, the causative factor of urotoxic side-effects of cyclophosphamide, ifosfamide, trofosfamide and sufosfamide. *Arzneimittel-Forschung*, **29**, 659–61.

Chak, L. Y., Sikic, B. I., Tucker, M. A., Horns, R. C., and Cox, R. S. (1984). Increased incidence of acute nonlymphocytic leukemia following therapy in patients with small cell carcinoma of the lung. *Journal of Clinical Oncology*, **2**, 385–90.

Coleman, C. N., Williams, C. J., Flint, A., Glatstein, E. J., Rosenberg, S. A., and Kaplan, H. S. (1977). Hematologic neoplasia in patients treated for Hodgkin's disease. *New England Journal of Medicine*, **297**, 1249–52.

Coleman, M., Easton, D. F., Horwich, A., and Peckham, M. J. (1988). Second malignancies and Hodgkin's disease—The Royal Marsden Hospital experience. *Radiotherapy and Oncology*, **11**, 229–38.

Coltman, C. A. and Dixon, D. O. (1982). Second malignancies complicating Hodgkin's disease: a Southwest Oncology Group 10-year followup. *Cancer Treatment Reports*, **66**, 1023–33.

Consensus Conference (1985). Adjuvant chemotherapy for breast cancer. *Journal of the American Medical Association*, **254**, 3461–3.

Curtis, R. E., Boice, J. D., Kleinerman, R. A., Flannery, J. T., and Fraumeni, J. F. (1985). Summary: multiple primary cancers in Connecticut, 1935–82. *National Cancer Institute Monographs*, **68**, 219–42.

Cuzick, J., Erskine, S., Edelman, D., and Galton, D. A. G. (1987). A comparison of the incidence of the myelodysplastic syndrome and acute myeloid leukaemia following melphalan and cyclophosphamide treatment for myelomatosis. *British Journal of Cancer*, **55**, 523–9.

de Vathaire, F., et al. (1989a). Long-term risk of second malignant neoplasm after a cancer in childhood. *British Journal of Cancer*, **59**, 448–52.

de Vathaire, F., et al (1989b). Role of radiotherapy and chemotherapy in the risk of second malignant neoplasms after cancer in childhood. *British Journal of Cancer*, **59**, 792–6.

Fishbein, L. (1987). Review—perspectives on occupational exposure to antineoplastic drugs. *Archiv für Geschwulstforschung*, **57**, 219–48.

Fisher, B., Rockette, H., Fisher, E. R., Wickerham, D. L., Redmond, C., and Brown, A. (1985). Leukemia in breast cancer patients following adjuvant chemotherapy or postoperative radiation: the NSABP experience: *Journal of Clinical Oncology*, **3**, 1640–58.

Glicksman, A. S., Pajak, T. F., Gottlieb, A., Nissen, N., Stutzman, L., and Cooper, M. R. (1982). Second malignant neoplasms in patients successfully treated for Hodgkin's disease: a cancer and leukemia group B study. *Cancer Treatment Reports*, **66**, 1035–44.

Greene, M. H., Young, R. C., Merrill, J. M., and DeVita, V. T. (1983). Evidence of a treatment dose-response in acute nonlymphocytic leukemias which occur after therapy of non-Hodgkin's lymphoma. *Cancer Research*, **43**, 1891–8.

Greene, M. H., et al. (1986). Melphalan may be a more potent leukemogen than cyclophosphamide. *Annals of Internal Medicine*, **105**, 360–7.

Haas, J. F., et al. (1987). Risk of leukaemia in ovarian tumour and breast cancer patients following treatment by cyclophosphamide. *British Journal of Cancer*, **55**, 213–18.

Habs, M. R. and Schmähl, D. (1983). Prevention of urinary bladder tumors in cyclophosphamide-treated rats by additional medication with the uroprotectors sodium 2-mercaptoethane sulfonate (mesna) and disodium 2,2'-dithio-*bis*-ethane sulfonate (dimesna). *Cancer*, **51**, 606–9.

Hawkins, M. M., Draper, G. J., and Kingston, J. E. (1987). Incidence of second primary tumours among childhood cancer survivors. *British Journal of Cancer*, **56**, 339–47.

Hemminki, K. and Ludlum, D. B. (1984). Covalent modifications of DNA by antineoplastic agents. *Journal of the National Cancer Institute*, **73**, 1021–8.

IARC (International Agency for Research on Cancer) (1987*a*). *Monographs on the evaluation of carcinogenic risks to humans*, Suppl. 7, *Overall evaluations of carcinogenicity: an updating of IARC Monographs Volumes 1–42*. IARC, Lyon.

IARC (International Agency for Research on Cancer) (1987*b*). *Monographs on the evaluation of carcinogenic risks to humans*, Suppl. 6, *Genetic and related effects: an updating of selected IARC Monographs from Volumes 1–42*. IARC, Lyon.

IARC (International Agency for Research on Cancer) (1990). *IARC Monographs on the evaluation of carcinogenic risks to humans*, Vol. 50, *Pharmaceutical drugs*. IARC, Lyon.

Johnson, D. H., Porter, L. L., List, A. F., Hande, K. R., Hainsworth, J. D., and Greco, J. A. (1986). Acute nonlymphocytic leukemia after treatment of small cell lung cancer. *American Journal of Medicine*, **81**, 962–8.

Kaldor, J. M., et al. (1987). Second malignancies following testicular cancer, ovarian cancer and Hodgkin's disease: an international collaborative study among cancer registries. *International Journal of Cancer*, **39**, 571–85.

Kaldor, J. M., Day, N. E., and Hemminki, K. (1988). Quantifying the carcinogenicity of antineoplastic drugs. *European Journal of Cancer and Clinical Oncology*, **24**, 703–11.

Kaldor, J. M., et al. (1990*a*). Leukemia following chemotherapy for ovarian cancer. *New England Journal of Medicine*, **322**, 1–6.

Kaldor, J. M., et al. (1990*b*). Leukemia following Hodgkin's disease. *New England Journal of Medicine*, **322**, 7–13.

Kempf, S. R. and Ivankovic, S. (1986*a*). Chemotherapy-induced malignancies in rats after treatment with cisplatin as single agent and in combination: preliminary results. *Oncology*, **43**, 187–91.

Kempf, S. R. and Ivankovic, S. (1986*b*). Carcinogenic effect of cisplatin (*cis*-diammine-dichloroplatinum(II), CDDP) in BD IX rats. *Journal of Cancer Research and Clinical Oncology*, **111**, 133–6.

Kinlen, L. J. (1985). Incidence of cancer in rheumatoid arthritis and other disorders after immunosuppressive treatment. *American Journal of Medicine*, **78**, 44–9.

Kinlen, L. J., Shiel, A. G. R., Peto, J., and Doll, R. (1979). Collaborative United Kingdom–Australasian study of cancer in patients treated with immunosuppressive drugs. *British Medical Journal*, **2**, 1461–6.

Lindahl, T., Sedgwick, B., Sekiguchi, M., and Nakabeppu, Y. (1988). Regulation and expression of the adaptive response to alkylating agents. *Annual Review of Biochemistry*, **57**, 133–57.

List, A. F., Doll, D. C., and Greco, F. A. (1985). Lung cancer in Hodgkin's disease: association with previous radiotherapy. *Journal of Clinical Oncology*, **3**, 215–21.

Muir, C., Waterhouse, J., Mack, T., Powell, J., and Whelan, S. (ed.) (1987). *Cancer Incidence in Five Continents, Volume V*, IARC Scientific Publications No. 88. International Agency for Research on Cancer, Lyon.

NCI (National Cancer Institute) (1988). *1987 Annual cancer statistics review, including cancer trends*: 1950–1985, NIH Publication No. 88–2789. US Department of Health and Human Services, Maryland.

Paukowits, W. R., Laerum, O. D., and Guignon, M. (1982). Isolation, characterisation and synthesis of a chalone-like haemoregulatory peptide. In *Biological regulation of cell proliferation*, (ed. R. Baserga), pp. 111–20. Raven Press, New York.

Pedersen-Bjergaard, J., *et al.* (1980). Acute non-lymphocytic leukemia in patients with ovarian carcinoma following long-term treatment with treosulfan (= dihydroxybusulfan). *Cancer*, **45**, 19–29.

Pedersen-Bjergaard, J., Østerlind, K., Hansen, M., Philip, P., Pedersen, A. G., and Hansen, H. H. (1985a). Acute nonlymphocytic leukemia, preleukemia, and solid tumors following intensive chemotherapy of small cell carcinoma of the lung. *Blood*, **66**, 1393–7.

Pedersen-Bjergaard, J., *et al.* (1985b). Risk of acute nonlymphocytic leukemia and preleukemia in patients treated with cyclophosphamide for non-Hodgkin's lymphomas. *Annals of Internal Medicine*, **103**, 195–200.

Pedersen-Bjergaard, J., Rørth, M., Avnstrøm S., Philip, P., and Hou-Jensen, K. (1985c). Acute nonlymphocytic leukemia following treatment of testicular cancer and gastric cancer with combination chemotherapy not including alkylating agents: report of two cases. *American Journal of Hematology*, **18**, 425–9.

Pedersen-Bjergaard, J., *et al.* (1987). Risk of therapy-related leukaemia and preleukaemia after Hodgkin's disease: relation to age, cumulative dose of alkylating agents, and time from chemotherapy. *Lancet*, **11**, 83–8.

Pedersen-Bjergaard, J., *et al.* (1988). Carcinoma of the urinary bladder after treatment with cyclophosphamide for non-Hodgkin's lymphoma. *New England Journal of Medicine*, **318**, 1028–32.

Rustin, G. J. S., Rustin, F:, Dent, J., Booth, M., Salt, S., and Bagshawe, K. D. (1983). No increase in second tumors after cytotoxic chemotherapy for gestational trophoblastic tumors. *New England Journal of Medicine*, **308**, 473–6.

Scheef, W., *et al.* (1979). Controlled clinical studies with an antidote against the urotoxicity of oxazaphosphorines: preliminary results. *Cancer Treatment Reports*, **63**, 501–5.

Seo, I. S., Clark, S. A., McGovern, F. D., Clark, D. L., and Johnson, E. H. (1985). Leiomyosarcoma of the urinary bladder 13 years after cyclophosphamide therapy for Hodgkin's disease. *Cancer*, **55**, 1597–603.

Spirtas, R. and Kaminski, R. (1978). Angiosarcoma of the liver in vinyl chloride/polyvinyl chloride workers. 1977 update of the NIOSH register. *Journal of Occupational Medicine*, **20**, 427–9.

Storm, H. H., *et al.* (1985). Summary: multiple primary cancers in Denmark, 1943–80. *National Cancer Institute Monographs*, **68**, 411–30.

Thiede, T. and Christensen, B. C. (1969). Bladder tumours induced by chlornaphazine: a five-year follow-up study of chlornaphazine-treated patients with polycythaemia. *Acta Medica Scandinavica*, **185**, 133–7.

Tucker, M. A., et al. (1987a). Leukemia after therapy with alkylating agents for childhood cancer. *Journal of the National Cancer Institute*, **78**, 459–64.

Tucker, M. A., et al. (1987b). Bone sarcomas linked to radiotherapy and chemotherapy in children. *New England Journal of Medicine*, **317**, 588–93.

Tucker, M. A., Coleman, C. N., Cox, R. S., Varghese, A., and Rosenberg, S. A. (1988). Risk of second cancers after treatment for Hodgkin's disease. *New England Journal of Medicine*, **318**, 76–81.

UNSCEAR (United Nations Scientific Committee on the Effects of Atomic Radiation) (1986). *Genetic and somatic effects of ionizing radiation*, Publication E.86.IX.9. UNSCEAR, New York.

Valagussa, P., et al. (1980). Second malignancies in Hodgkin's disease: a complication of certain forms of treatment. *British Medical Journal*, **280**, 216–19.

Valagussa, P., Santoro, A., Fossati-Bellani, F., Banfi, A., and Bonadonna, G. (1986). Second acute leukemia and other malignancies following treatment for Hodgkin's disease. *Journal of Clinical Oncology*, **4**, 830–7.

Wall, R. L. and Clausen, K. P. (1975). Carcinoma of the urinary bladder in patients receiving cyclophosphamide. *New England Journal of Medicine*, **293**, 271–3.

5
Chemotherapy and immunosuppression for non-malignant conditions

MARK S. DORREEN and BARRY W. HANCOCK

General introduction

Immunosuppressive and cytotoxic drugs have been used for many years in a wide variety of benign conditions. Corticosteroids have been used very widely, but other classes of drugs have also been used (Table 5.1), particularly the antimetabolites (methotrexate and azathioprine) and the alkylating drugs (principally cyclophosphamide, but also chlorambucil, melphalan, thiotepa, and mustine).

In organ transplantation, azathioprine, prednisolone and, more recently, cyclosporin A are used to induce a deliberate state of immunosuppression in order to prevent graft rejection. Although their place in the treatment of other conditions is not clearly established, cytotoxic and immunosuppressive drugs have also been used in a wide range of inflammatory disorders, including rheumatoid arthritis, the collagen vascular diseases, inflammatory bowel disease, chronic hepatitis, chronic glomerulonephritis, and other diseases characterized as predominantly auto-immune in nature. However, it is in the rheumatic diseases that most experience has been gained and that evidence for their efficacy is strongest.

Cytotoxic drugs are indicated, particularly, in the management either of severe disease unresponsive to moderate doses of prednisolone, or else where corticosteroid therapy has resulted in unacceptable side effects. Even where corticosteroids need to be maintained, these can usually be continued in lower doses because of the 'steroid-sparing' properties of cytotoxic drugs. However, cytotoxic drugs are also directly toxic, particularly to the gastro-intestinal system and the bone marrow. Such treatment, therefore, requires careful monitoring although the acute effects can often be overcome through careful dosage adjustment.

In contrast, the long-term toxicity of immunosuppressive/cytotoxic therapy has posed the most serious and potentially life-threatening problems to patients receiving such treatment. Among the long-term complications, the occurrence of malignancy has been increasingly recognized as a cause of morbidity and mortality. The development of possible treatment-related tumours has led some centres to question or reappraise the indications for the use of immunosuppressive/cytotoxic drugs in non-malignant disease.

Table 5.1 Immunosuppressive drugs commonly used in benign disease

	Indications
Alkylating drugs	
Cyclophosphamide	Systemic lupus erythematosus
	Rheumatoid arthritis
	Wegener's granuloma
	Other chronic inflammatory disorders
Chlorambucil	Similar indications but less commonly used
Melphalan	
Antimetabolites	
Azathioprine	Suppression of organ allograft rejection
	Systemic lupus erythematosus
	Rheumatoid arthritis
	Wegener's granuloma and other chronic inflammatory disorders
Methotrexate	Severe, unremitting psoriasis with systemic involvement
Corticosteroids	
Prednisone	Suppression of organ allograft rejection
Prednisolone	Systemic lupus erythematosus
	Rheumatoid arthritis
	A wide range of other disorders
Cyclosporin A	Suppression of organ allograft rejection by inhibition of lymphocyte proliferation

This question and the magnitude of the problem faced by the clinician is addressed below, in reviews and analysis of the many reports and studies which have examined the occurrence of neoplasia during or after treatment for non-malignant disease.

Organ transplantation

By 1970 it was recognized that renal transplant recipients were at a substantially increased risk from non-Hodgkin's lymphoma—particularly 'reticulum cell sarcomas' with a predilection for the brain (Penn *et al.* 1969; Doll and Kinlen 1970). A collaborative UK–Australasian study of such patients (treated with azathioprine, cyclophosphamide or chlorambucil) showed a 60-fold increase of non-Hodgkin's lymphoma together with an excess of squamous cell skin cancer and mesenchymal tumours (Kinlen *et al.* 1979). Since then there have been many reports of an increased risk of cancer after organ transplantation (for reviews see Sheil 1986; Penn 1988*b*).

The most commonly described tumours are skin carcinomas and lymphomas. The rates of cancer incidence have, however, been assessed in many different ways making interpretation of data very difficult, and large, controlled studies are lacking. In some studies the data collected exclude skin carcinomas. Other groups have tried, albeit subjectively, to distinguish between malignancy which might have been induced by immunosuppressive treatment and other malignant disease (Brunner *et al.* 1989). It may be that the relative risks of cancer are small (MacLeod and Catto 1988); certainly death rates from malignant disease are less than from cardiovascular causes (McGeown *et al.* 1988; Brunner *et al.* 1989). However, most authorities believe that the overall risk of cancer (of any type) in patients being followed up after organ transplant is up to 10 per cent, skin cancer being commonest and malignant lymphoma following closely.

Aetiology

The mechanism by which the incidence of such a narrow range of tumours could be increased is uncertain. There seems little doubt that immunosuppression must have a major role, and the risk may be greatest (particularly for malignant lymphoma) when high doses of immunosuppressive drugs are given (Bieber *et al.* 1981; Penn 1982; Kinlen *et al.* 1983). Other mechanisms of oncogenesis may be acting, either singly or in combination. Cytogenetic abnormalities induced by extrinsic factors such as radiation, chemicals or viruses may lead to the development of malignant clones of cells. Rapidly dividing cells would be most at risk, which might explain the predominance of lymphoreticular tumours. Chronic antigen stimulation of the immune system can cause lymphoproliferation which may become neoplastic. After transplantation this is likely to involve the host versus graft reaction. Evidence for the effect of oncogenic viruses is becoming increasingly apparent (Hanto *et al.* 1985; Sheil 1986). Infection with Epstein–Barr virus may cause lymphoproliferative disease and it is possible that other viruses linked with human tumours could also be important agents.

Immunosurveillance

The association between cancer and immunosuppression has long been recognized. According to the immunosurveillance theory, defective immunological control of potentially neoplastic cells allows their proliferation and development into cancer (Burnet 1971). Immunosurveillance remains a controversial subject, however (Penn 1981), and many aspects of the original theory have been discredited, although biological processes (some immune-mediated) undoubtedly have a role in carcinogenesis (for review see Hancock and Ward 1985). The early theories were based on lymphocyte-mediated mechanisms, which would eliminate newly arising malignant cells; impairment of immunity would then lead to increased malignancy. Two findings which do not support the concept of immunological surveillance are:

1. The increase in malignancy is restricted to certain types of cancer, rather than the broad spectrum that might be expected (in approximate proportion to the usual distribution of malignancy) if a general mechanism of immunosurveillance were involved.
2. Certain congenital immunodeficiency states, e.g. chronic granulomatous disease and the Di George syndrome, which involve the T-cells, phagocytes, and natural killer (NK) cells that are fundamental to the proposed immunological surveillance mechanism, are not necessarily associated with an increased risk of developing malignancy.

Latency

It is a feature of post-transplant cancers that they have a relatively short latency between the start of immunosuppression and tumour development. If immunosuppressive drugs were actually carcinogenic, the latent period would be expected to be much longer and the spectrum of tumours different.

Viruses

Viruses may have an important part to play in the genesis of cancer in transplant recipients. Squamous cell carcinoma of the skin occurs more often than basal cell carcinoma in post-transplant patients, the reverse of the normal situation. There appears to be a definite association with viruses when squamous cell carcinoma develops in association with warts, and of great interest is the recent discovery of the papilloma virus genome in at least one-third of post-transplant squamous skin carcinomas (Sheil 1986).

There may be an increased incidence of cervical cancer in immunosuppressed transplant recipients; these findings have been related to human papilloma virus and also to herpes simplex virus. In one study, abnormal cytology was found in approximately two-thirds of such patients, and human papilloma viruses believed to be of malignant potential were isolated from about half of these (Alloub *et al.* 1988). Kaposi's sarcoma and lymphoma in the central nervous system seem to be unequivocally related to post-transplant immunosuppression (a situation mirrored in Human Immunodeficiency Virus-induced AIDS); cytomegalovirus may also have an aetiological role in Kaposi's sarcoma.

Of by far the greatest interest, however, is the recognized association between the Epstein–Barr virus (EBV) and lymphoreticular malignancies (Hanto et al. 1983). EBV has been particularly linked to primary central nervous system (CNS) lymphoma in transplant recipients (Hatchberg et al. 1983). Pharyngeal shedding of the virus occurs more frequently in transplant patients as compared to normal sero-positive subjects (Chang et al. 1978) and there are also substantial rises in antibody titre following transplantation (Cheeseman et al. 1980). It has been suggested that immunosuppressive agents inhibit T-cell function, thus allowing unrestrained proliferation of B-cells in response to primary or latent viral infection, and that this polyclonal proliferation ultimately develops into a monoclonal B-cell lesion, i.e. lymphoma (Klein and Purtillo 1981).

The immunological responses to EBV are complex. The mechanisms for control of EBV-infected B-lymphocytes include:

(1) production of specific cytotoxic T-cells;
(2) initiation of natural killer (NK) cell immune surveillance;
(3) initiation of EBV-specific suppressor T-cells;
(4) production of specific antibodies against the viral capsid and nuclear antigens;
(5) the production by infected B-cells of cytokines such as interferon, blocking viral replication.

Many of these mechanisms are impaired in immunosuppressed transplant recipients; antibody responses are inappropriate, there is a reduction in the number of specific cytotoxic T-cells (Crawford et al. 1981), and NK activity is impaired (Lipinski et al. 1980)—leading to impairment of the necessary immune interactions (for review see Hanto et al. 1985).

The exact mechanism of malignant transformation from polyclonal to monoclonal B-cell proliferation is still unknown (Hanto et al. 1983) although a cytogenetic effect similar to that seen in Burkitt's lymphoma seems the most plausible explanation. Much of the work on viruses and lymphoreticular proliferative disorders was done when cyclosporin A first came into clinical use; lymphomas seemed especially likely when the drug was given in the original high dosage or in combination with other immunosuppressive agents (Penn 1988a). With current dosage regimens it does seem that the incidence of lymphoma and other tumours in patients treated with cyclosporin A is no greater than in patients treated with more conventional 'immunosuppressive' regimens (e.g. azathioprine and prednisolone) (Beveridge 1983). Penn (1988b), however, remains cautious, suggesting that longer follow-up of large numbers of patients will be necessary to clarify the full carcinogenic potential of cyclosporin A. The International Agency for Research on Cancer Monograph Working Group (IARC, 1990) has recently reviewed all available evidence and concludes that the incidence of lymphoma and probably Kaposi's sarcoma is remarkably high in patients treated with cyclosporin; they also comment that it is noteworthy that some lymphomas have regressed following discontinuation of therapy and that current evidence is consistent with the hypothesis that intensity of immunosuppression is an important feature in lymphoma incidence. It is unclear why about half the lymphomas seen in transplant recipients are of primary CNS origin. The most likely explanation is that the CNS is an immunologically privileged site, so that any virus gaining entry to the CNS would be particularly likely to escape immune control when immunosuppressive therapy is being given.

Specific cancers after organ transplantation

A résumé of information concerning the cancers seen after transplantation is presented in Table 5.2, derived from transplant registers around the world. The patterns of increased risk are unusual, and specific organs for which cancer risk is most increased are briefly discussed below.

Skin carcinoma There is considerable geographical variation in the incidence of many cancers seen after transplantation, and this is especially so for skin carcinoma, where other geographically variable risk factors are also important—high sunlight exposure (Hardie *et al.* 1980); Anglo-Saxon or Celtic origin; fair skin, blue eyes, and light sensitivity (Kelly *et al.* 1987)—in fact, risk factors similar to those in the normal population. There is an increased risk of cancer with length of time since transplantation; this is particularly so for skin carcinoma. In patients followed for up to 15 years the cumulative incidence of skin carcinoma was 44 per cent, and 14 per cent for other cancers (Sheil 1986). In terms of mortality, skin carcinoma does not figure highly, since the prognosis with treatment is still good. The risk for melanoma is probably the same, or only minimally greater, than that seen in the general population.

Lymphoma All types of lymphoid malignancy, the majority of these being non-Hodgkin's lymphomas, have been described in transplant recipients, with an incidence estimated variously at between 10- and 50-fold above that expected (Hoover and Fraumeni 1973; Kinlen *et al.* 1979; Sheil 1986). Lymphomas account for between 12 and 50 per cent of tumours (the latter excluding skin carcinomas), compared with 3 per cent or less in the general population; the incidence is greater in older transplant patients (Brunner *et al.* 1989) and in patients undergoing cardiac (Bieber *et al.* 1981) or hepatic transplant (Penn 1982), where immunosuppression is generally more prolonged and severe than with renal transplantation (Kinlen *et al.* 1983). About half are primary cerebral lymphomas (Patchell 1988; Dorreen *et al.* 1988). These tumours, variously termed microglioma and reticulum cell sarcoma, are now acknowledged to be true lymphomas, indistinguishable from non-Hodgkin's lymphomas arising in other extra-

Chemotherapy, immunosuppression for non-malignant conditions 77

Table 5.2 Distinctive features of post-transplant tumours compared to tumours in the general population: résumé of data from various transplant registries

	Proportion of post-transplant tumours (%)	Risk	Latency	Histology	Clinical	Prognosis	Role of immunosuppression
Skin (excl. melanoma)	c.40	8-fold to 20-fold	Increasing with time post-transplant	Squamous > basal cell	Lymphatic metastases and multiple lesions more common	Less favourable	Moderate
Lymphoma	10–15	Over 10-fold (over 1000-fold for CNS)	Short	Usually high grade	Extranodal (particularly brain) sites more common in older patients	Poor	Major (especially CNS)
Kaposi's sarcoma	5–10	Over 1000-fold	Short	Unremarkable	Reduced male/female ratio; more aggressive	Poor	Major
Cervix Colon Kidney Leukaemia Thyroid	c.15	2-fold to 5-fold	Relatively short	Invasive lesions of cervix more common	Cervix cancer occurs early	Less favourable	Minimal except for cervix
Breast Bronchus Prostate Melanoma	25	None	Similar	Unremarkable	Unremarkable	Similar	None

nodal sites. Most of them are of B-cell origin and are high grade (unfavourable prognosis) in their morphology and behaviour.

Kaposi's sarcoma This tumour, extremely rare in Western populations before the 1980s, when the AIDS epidemic was identified, accounts for up to 10 per cent of cancers seen following transplantation, an estimated 500- to 1000-fold risk over normal (Harwood *et al.* 1979; Sheil 1986) and about one-third of patients have visceral as opposed to skin or mucosal disease.

Other tumours Virtually every type of cancer has been reported to occur with increased frequency, reduced latency and in a younger age range; this includes leukaemia and carcinomas of the uterine cervix, vulva, perineum, colon, kidney, thyroid, and liver (Kinlen *et al.* 1979; Birkeland 1983; Sheil 1986; Penn 1988*b*).

However, in the European Dialysis and Transplant Association (EDTA) study (Brunner *et al.* 1989) when lymphoma, skin carcinoma, Kaposi's sarcoma, cervix, and colon carcinoma were excluded from analysis, other tumours, including melanoma and the more common bronchial and breast tumours, were recorded with an incidence mostly within the range given by cancer registries for the general population.

Transplantation after cancer

Undoubtedly, a history of cancer before transplantation predisposes to cancer following the procedure (Penn 1983*a*). Patients with malignancies involving organs to be replaced at transplantation are at particularly high risk for developing recurrence of the original cancer. Patients with previous malignancy of tissues not involved in transplantation are not only at risk of new malignancy (see above) but also at increased risk of recurrence of their original malignancy. It seems that patients treated for their first cancer at least 2 years before the transplantation procedure are less likely to show recurrence of the original lesion.

Prognosis

The behaviour of tumours seen in transplant recipients reflects that seen with all immunosuppressed patients, in that they are clinically more aggressive and unpredictable. This is reflected in their poor response to treatment and, ultimately, in their generally poor prognosis. A major problem in treating tumours in this group is that the underlying immunosuppressed status limits the available therapeutic options, particularly for intensive chemotherapy.

The EDTA study (Brunner *et al.* 1989), however, has shown that survival from cancers occurring in grafted patients with chronic renal failure is no worse than that seen for non-grafted patients, despite the immunosuppressive therapy which many transplanted patients continued to receive during treatment of their malignant tumours.

The theoretical promise held out by the use of biological response modifier therapy with cytokines such as interferon and interleukin-2 has not yet proved useful in clinical practice, although in some early polyclonal lymphoproliferative disorders, including those of the central nervous system, reversal of the immunosuppression has led to regression of the lesion (Starzl *et al.* 1984). Another potential line of attack would be the use of anti-viral agents, or indeed pre-transplant vaccination, to reduce the risk of viral oncogenesis. However, the role of vaccination might be limited by the poor immune response associated with the underlying illness (for example uraemia) and as yet, the use of anti-viral agents (such as zidovudine), has not been shown to reduce tumour incidence.

Malignancy in other benign diseases

The risk of malignancy in patients with other benign disorders is considerably less clear than in transplant recipients, and much of the evidence is based on case reports. However, it has long been apparent that some inherited immunodeficiency diseases, such as ataxia telangiectasia and the Wiskott–Aldrich syndrome, are associated with an increased risk of malignancy, particularly malignant lymphoma (Heidelberger and LeGolvan 1974; Frizzera *et al.* 1980; Penn 1986). AIDS-related malignant lymphoma is also the subject of a rapidly accumulating and extensive body of literature but, as in organ transplantation, there is a predilection for such tumours to present in the central nervous system (Snider *et al.* 1983; Gill *et al.* 1985; Helweg-Larsen *et al.* 1986).

The scope of the problem of malignancy in autoimmune disease remains ill-defined. It would seem reasonable to suppose that many of the potentially oncogenic factors which are produced deliberately during organ transplantation in order to avoid graft rejection may also apply in endogenous autoimmune disease. These factors would include loss of immune surveillance, impaired immunoregulation, and chronic antigenic stimulation. Certain tumours probably also arise as a direct result of the carcinogenic properties of particular drugs or their metabolites (Schmähl 1977; Rieche 1984; Kaldor and Lasset, Chapter 4, this volume).

Among individual drugs used for non-malignant disease, cyclophosphamide has been implicated most consistently in the development of malignancy (Schmähl *et al.* 1982). The role of azathioprine is more equivocal, and it is normally used in lower doses and for considerably shorter periods of time than when following organ transplantation. In a brief follow-up to the UK–Australasian study referred to earlier (Kinlen et al. 1979), the incidence of tumours other than lymphoma in patients who had received azathioprine but had not had a transplant was no higher than would have been expected in the general population.

Lymphoma

Predisposition to lymphoma in rheumatic disease Rheumatoid arthritis is a disease characterized by intense immunological activity and patients may present with reactive lymphadenopathy and, less commonly, splenomegaly (Motulsky et al. 1952). Both reactive germinal B-cell follicles and diffuse infiltrates of T-cells are also known to occur in the synovial membrane as part of the rheumatoid process (Young et al. 1984; Vaughan 1985). On these grounds alone, it might well be expected that the incidence of malignant lymphoma would be higher in rheumatoid arthritis, even in the absence of therapy with potentially carcinogenic drugs.

Lea (1964) reported a highly significant association between rheumatic disease and the occurrence of both lymphoma and leukaemia. The patient groups were only loosely defined, however, and the study was not analysed for the effects of previous therapy. In a later study of nearly 3000 patients with rheumatoid arthritis, followed up at the University of Michigan, Lewis et al. (1976) reported 17 malignancies (including two lymphomas), an incidence of 0.6 per cent, which was significantly lower than expected.

A number of recent studies, however, now suggest that the incidence of lymphoid malignancies is significantly increased in rheumatoid arthritis. Using Finnish cancer registry data, Isomaki et al. (1982) reported a highly significant increase in the occurrence of malignant lymphoma in a population of more than 46 000 patients with rheumatoid arthritis. As in the study of Lea (1964), however, the analysis did not exclude a possible effect of previous therapy. A statistically significant 8-fold excess of lymphoma has been reported in patients with rheumatoid arthritis in Birmingham, UK (Prior et al. 1984; Prior 1985). The incidence of lymphoma increased in a linear manner among patients who had been followed up for at least 5 years, resulting in cumulative risk estimates of 10 per cent at 10 years after diagnosis and 20 per cent after 20 years. In a retrospective analysis of lymphoid malignancies in rheumatoid patients, almost 75 per cent were B-cell disorders, the majority non-Hodgkin's lymphoma (Symmons et al. 1984; Symmons 1985). None of the patients described in the Birmingham studies had received immunosuppressive or cytotoxic therapy, although most had received phenylbutazone and penicillamine.

Although the majority of lymphomas associated with rheumatoid arthritis have proved to be B-cell tumours, Jack et al. (1986) reported four cases of malignant histiocytosis in patients with proven rheumatoid disease. While the precise histogenesis of these tumours remains uncertain, they behave as tumours of neoplastic macrophages and should probably not be regarded as true lymphomas.

In Sjögren's syndrome, which may accompany rheumatoid disease, the association with non-Hodgkin's lymphoma now appears to be well established (Talal and Bunim 1964; Talal *et al.* 1967; Anderson and Talal 1971; Whaley *et al.* 1973). Kassan *et al.* (1978) have reported a 44-fold excess of lymphoma, enhanced still further through prior treatment with irradiation or chemotherapy.

The occurrence of lymphoma has been most striking in patients with severe disease affecting extra-salivary tissue, notably lymph node and spleen (Talal and Bunim 1964; Talal *et al.* 1967; Anderson and Talal 1971). Pathologically, a spectrum of lymphoid abnormalities has been observed, up to and including lymphoma. Most lymphomas are high-grade malignancies, although examples of Waldenström's macroglobulinaemia have been observed (Anderson and Talal 1971). Zulman *et al.* (1978) have demonstrated monoclonal light chain restriction within these lymphomas.

The relationship between other connective tissue disorders and malignant lymphoma is uncertain. Canoso and Cohen (1974) reported eight tumours, including one non-Hodgkin's lymphoma, among 70 patients with systemic lupus erythematosus (SLE). Lewis *et al.* (1976) reported an overall increase in the incidence of malignancy among nearly 500 patients with SLE, although the only case of lymphoma was a patient with Hodgkin's disease. There have, however, been consistent reports of a possible association of SLE with Hodgkin's disease (Cammarata *et al.* 1963; Andreev and Zlatkov 1968; Cudworth and Ellis 1972; Green *et al.* 1978; Cohen *et al.* 1988), although the occurrence of non-Hodgkin's lymphomas, usually low-grade B-cell tumours, has also been noted in these studies.

Despite this suggestion of a possible predisposition to lymphoma in SLE, large-scale data are lacking. The only exception has been the largely negative University of Michigan study (Lewis *et al.* 1976), although the diagnosis of one case of Hodgkin's disease, among 18 tumours observed, is of some interest. There is, however, an important caveat in seeking to ascribe an association between the collagenoses and malignant lymphoma, since lymphomas may present clinically with paraneoplastic syndromes closely resembling collagen vascular disorders. Examples include dermatomyositis (Calabro 1967) and polyarteritis nodosa (Miller 1967). More recently, an association between hairy cell leukaemia, a B-cell lymphoproliferative malignancy, and a polyarteritis-like syndrome, has been described (Elkon *et al.* 1979; Goedert *et al.* 1981). Elkon and co-workers attributed the latter to a serum-sickness type of reaction resulting from the defective clearance of immune complexes, inherent to the primary disease process.

Lymphoma following immunosuppressive therapy in rheumatic disease
Lipsmeyer (1972) reported a cerebral lymphoma in a patient who had received azathioprine for SLE. However, exposure to the drug had been relatively brief (12 months) and the possibility that the cerebral tumour was a complication of the disease process rather than its treatment cannot be excluded from such a case report. As yet, the relationship between lymphoma and prior immunosuppressive therapy remains equivocal, although several studies indicate no clear evidence of such an association (Farber *et al.* 1979; Lewis *et al.* 1980; Boyle *et al.* 1981; Hazleman and De Silva 1982; Kirsner *et al.* 1982; Speerstra *et al.* 1982; Wessel *et al.* 1988).

Fosdick *et al.* (1969), however, reported two lymphoid malignancies in 108 patients with rheumatoid arthritis treated with cyclophosphamide. The same group later described five cases of non-Hodgkin's lymphoma in over 400 similarly-treated patients (Love and Sowa 1975). Baltus *et al.* (1983) found a 4-fold increase in malignancies of all types in rheumatoid patients treated with cyclophosphamide, as compared both with untreated patients and the general population. Three of the 15 tumours diagnosed in the cyclophosphamide-treated group were non-Hodgkin's lymphomas, whereas no lymphomas were noted among four tumours seen in the control group of rheumatoid patients. The peak incidence for the occurrence of all tumours was in the sixth to ninth years of follow-up. Among patients who developed a tumour, the mean total dose of cyclophosphamide received was 82 g, compared with 61 g among those who did not ($p = 0.054$). Baker *et al.* (1987) observed 37 tumours (including two lymphomas) among 119 patients with rheumatoid arthritis treated with cyclophosphamide, compared with 16 tumours (no lymphoma) in a matched control group of rheumatoid patients. The mean total dose of 75 g cyclophosphamide among patients who developed a tumour was significantly higher than the mean dose of 46 g in those who did not.

Kinlen *et al.* (1979) reported an 11-fold excess of non-Hodgkin's lymphoma in non-transplant patients treated with either azathioprine, cyclophosphamide or chlorambucil, and the same order of risk was also observed in patients with rheumatoid arthritis and a variety of other collagen vascular disorders who had received similar treatment (Kinlen 1985). Pitt *et al.* (1987) reported three cases of aggressive non-Hodgkin's lymphoma in 41 rheumatoid patients treated with azathioprine, all of whom had received large cumulative doses over several years. Silman *et al.* (1988) observed four malignant lymphomas in 202 patients who had received azathioprine for rheumatoid arthritis, compared to two cases among 202 patients treated without this drug. The average daily dose of azathioprine was 300 mg, i.e. from two to three times the conventional dose schedule. Compared to the general population, these cases represented a 10-fold and 5-fold relative risk of lymphoma in the azathioprine and control groups, respectively. It was concluded that, in a disease with an inherent predisposition to lymphoid malignancy, the additional contribution to the risk of lymphoma from conventional treatment with azathioprine was likely to be small.

In conclusion, evidence that immunosuppressive/cytotoxic therapy enhances the occurrence of malignant lymphoma in rheumatic disease is provided by a number of studies, as summarized in Table 5.3, although the large number of negative studies also shows that the argument remains unresolved. Any excess risk resulting from the use of these drugs also needs to be set against a possible predisposition to lymphoma, which may be inherent to rheumatoid arthritis and the collagenoses. Finally, it remains unclear whether one class of drugs carries a greater potential for inducing lymphomatous transformation than any other. While few direct comparisons have been made, Kinlen (1985) demonstrated no obvious differences between azathioprine and the alkylating drugs.

Lymphoma in non-rheumatic disease There are few data on the occurrence of malignant lymphoma in other, non-malignant diseases, although Kinlen's survey (Kinlen 1985) covered a wide range of non-rheumatic conditions, including renal and inflammatory bowel disease, among which the excess risk of treatment-related lymphoma was considered to apply along with rheumatoid arthritis. Other data are restricted to case reports from which it is difficult to assess the risk with any certainty (Sharpstone *et al.* 1969; Chaplin 1982).

Acute myelogenous leukaemia

Acute myelogenous leukaemia following alkylating agent therapy for malignant disease is well recognized, and is discussed in Chapter 4 (Kaldor and Lasset, this volume). Leukaemic change appears to arise as a direct result of prolonged exposure of the bone marrow to alkylating drugs. Preleukaemic phases of up to several years' duration are well recognized and are characterized by an increasingly dysmyelopoietic bone marrow with karyotypic abnormalities most frequently involving translocations of chromosomes 5 and 7 (Cadman *et al.* 1977; Rowley *et al.* 1981; Pedersen-Bjergaard *et al.* 1987).

The occurrence of AML following the use of alkylating drugs in benign disease is now increasingly recognized (Love and Sowa 1975; Roberts and Bell 1976; Seidenfeld *et al.* 1976; De Bock and Peetermans 1977; Kapadia and Kaplan 1978; Hochberg and Shulman 1978; Grünwald and Rosner 1979; Kahn *et al.* 1979; Sheibani *et al.* 1980; Lebranchu *et al.* 1980; Baker *et al.* 1987; Gibbons and Westerman 1988). However, most of the foregoing are case reports or collections of case reports and there are insufficient data to estimate the relative or cumulative risk of alkylating drug therapy in benign disease. Kahn *et al.* (1979) observed an incidence of AML of just under 1 per cent in over 1800 patients treated with cyclophosphamide or

Table 5.3 Studies indicating an increased of lymphoma in rheumatic disease[1]

Group	Year	Study population[2]	Source	All tumours	Lymphoma	Lymphoma Obs/Exp	Details of therapy	Comments
Isomaki et al.	1982	46101	Finland	1202	57 (38: NHL) (19: HD)	2.7 $p<0.001$	Not stated	Inherent tendency to lymphoma suggested
Baltus et al.	1983	81 CP-treated 81 controls	Arnhem and Rotterdam	15 4	3: NHL 0		Mean total dose CP: 82 g in tumour pts; 61 g in others	
Prior, Symmons et al.	1984, 1985	489	Birmingham	42	10 (7: NHL) (1: CLL) (2: HD)	11 $p<0.001$	None received azathioprine or cyclophosphamide	Linear rise in risk of lymphoma in pts followed for 5 years
Kinlen	1985	1634: RA, SLE and other diseases	UK	59	6: NHL	11	68%: received CP 28%: received azathioprine 4%: received chlorambucil	
Pitt et al.	1987	41	London	Not stated	3: NHL		All received azathioprine	'Striking increase in incidence of lymphoma'
Baker et al.	1987	119 CP-treated 119 controls	Pittsburgh	37 16	2: NHL 0		Mean total dose CP: 75 g in tumour pts; 46 g in others	Incidence not significant but suggestive of increase
Silman et al.	1988	202 Aza-treated 202 controls	Cambridge	35 23	4: NHL 2: NHL	10 5	Median daily dose Aza 300 mg for 3 years	Inherent tendency to lymphoma; risk doubled with treatment

[1] Obs/Exp, Observed/expected ratio; RA, rheumatoid arthritis; NHL, non-Hodgkin's lymphoma; HD, Hodgkin's disease; CLL, chronic lymphatic leukaemia; CP, cyclophosphamide; Aza, azathioprine; pts, patients.
[2] All studies of patients with RA, except Kinlen (1985).

chlorambucil for rheumatoid arthritis. No instance of AML was noted in patients treated for less than 6 months or who had received a total of less than 1 g of chlorambucil or 50 g of cyclophosphamide. In five separate reports implicating cyclophosphamide in the occurrence of AML (Love and Sowa 1975; Roberts and Bell 1976; Kapadia and Kaplan 1978; Hochberg and Shulman 1978; Baker *et al*. 1987), the average total dose received was more than 100 g, given over several years.

While treatment with alkylating agents is clearly linked to AML, can a possible leukaemogenic effect of azathioprine be dismissed altogether? Certainly, cases of AML have been reported in patients treated with azathioprine for a variety of disorders (Silvergleid and Schrier 1974; Grünwald and Rosner 1979) and Urowitz *et al*. (1981) have described karyotypic abnormalities in association with azathioprine therapy in 78 per cent of bone marrows examined. Nonetheless, if there were a directly leukaemogenic effect of azathioprine, it would be expected that AML should be a major cause of mortality in organ allograft recipients, who are exposed to prolonged periods of treatment with large doses of azathioprine, and this is not the case.

Bladder cancer

It has long been recognized that treatment with cyclophosphamide carries a significant risk of urothelial toxicity, the most common manifestation of which is the occurrence of a haemorrhagic cystitis, usually reversible on cessation of treatment. Forni *et al*. (1964) described marked cellular atypia on urinary cytology in patients exposed to cyclophosphamide in the treatment of malignant lymphoma. However, it was not until the following decade that the association between bladder cancer and cyclophosphamide was recognized (Worth 1971). Subsequently, reports of bladder cancer following cyclophosphamide treatment for malignant disease appeared regularly (Dale and Smith 1974; Wall and Clausen 1975; Richtsmeier 1975; Ansell and Castro 1975; Seltzer *et al*. 1978; Fairchild *et al*. 1979; Fuchs *et al*. 1981). The length of exposure to the drug and the total dose received are both important factors, and most patients in these studies had received well in excess of 100 g of cyclophosphamide. In most cases treatment had also been complicated by the occurrence of at least one episode of haemorrhagic cystitis. Almost all cases have proved, histologically, to be transitional cell carcinomas.

Cox (1979*a*,*b*) described the aetiological role of acrolein, a urinary metabolite of cyclophosphamide, in causing a chemical cystitis. Prevention of the cystitis could be effected through prior treatment with the urinary protective agent, *N*-acetyl-cysteine. Cox proposed that, following the acute superficial necrosis of the bladder epithelium induced by acrolein, a rapid reparative phase follows, during which the actively regenerating cells are exposed to the full carcinogenic effects of either unchanged cyclophosphamide or one of its other metabolites, a hypothesis which has not been seriously challenged. While the bladder, in its function as a urinary reservoir, receives the major impact of cyclophosphamide toxicity, the whole of the urothelium is at risk, and Fuchs et al. (1981) described two transitional cell carcinomas of the renal pelvis following cyclophosphamide therapy.

With the increasing use of cyclophosphamide in benign disease, there have now been many reports of bladder cancer after prolonged treatment. As with the bladder tumours arising in patients with a previous malignancy, the majority of these tumours have been grade III (poorly-differentiated) transitional cell carcinomas, with a tendency to be locally extensive or metastatic. They have been described following cyclophosphamide treatment in rheumatoid arthritis (Plotz et al. 1979; Lewis et al. 1980; Garvin and Ball 1981; Hansen 1983; Ansher et al. 1983; Kinlen 1985; Baker et al. 1987), systemic lupus erythematosus (Plotz et al. 1979; Elliot et al. 1982), Wegener's granulomatosis (Stillwell et al. 1988), and a wide variety of other systemic inflammatory disorders (Kinlen et al. 1981; Schiff et al. 1982; Kinlen 1985). There has also been one report of bladder tumours in patients who had received large cumulative doses of azathioprine for ulcerative colitis (Scharf et al. 1977). No similar cases have been reported since, and the causal relationship of prior treatment with azathioprine, if any, remains speculative.

Estimates of the extent of the risk of bladder cancer remain incomplete, although Fairchild et al. (1979) calculated a 9-fold excess risk in cyclophosphamide-treated patients, based on extensive cancer registry data, while Kinlen and co-workers have quoted relative risks of between 10 (Kinlen et al. 1981) and 4 (Kinlen 1985). Baker et al. (1987) reported six bladder cancers in 119 rheumatoid patients treated with cyclophosphamide, an incidence of just under 5 per cent. No bladder tumours were observed in the control population of patients with rheumatoid arthritis. Stillwell et al. (1988) reported three bladder cancers in 111 patients who received cyclophosphamide for Wegener's granuloma, an incidence of around 3 per cent. In this context, therefore, the risk of cyclophosphamide-induced bladder cancer is likely to be small, but caution in the use of cyclophosphamide in non-malignant disease is still advisable, and it should probably be abandoned in all patients who develop haemorrhagic cystitis.

Kaposi's sarcoma

Although less striking in incidence than in transplant recipients, Kaposi's sarcoma (KS) is now well recognized as occurring with increased frequency in patients receiving immunosuppressive therapy (Penn 1983b). Cases have been reported among patients with rheumatoid arthritis (Gange and Jones 1978; Schottstaedt et al. 1987; Weiner et al. 1988), SLE (Klein et al 1974), polymyositis and dermatomyositis (Dantzig 1974; Weiss and Serushan 1982), Wegener's granuloma (Erban and Sokas 1988), temporal arteritis (Leung et al. 1981), bullous pemphigoid (Tye 1970; Scaparro et al. 1984), pemphigus vulgaris (Gange and Jones 1978), immune thrombocytopenia (Turnbull and Almeyda 1970), asthma (Hoshaw and Schwartz 1980), and myasthenia gravis (Snyder and Schwartz 1982). The great majority of patients had received prolonged courses of treatment with corticosteroids alone, although some had also been treated with either azathioprine or cyclophosphamide. Erban and Sokas (1988) consider the long-term use of corticosteroids to be the single most important aetiological factor in the occurrence of KS. In their own example, a 78-year-old man with Wegener's granuloma who had received treatment with both cyclophosphamide and prednisolone, there was complete regression of the nodules of KS with discontinuation of prednisolone, despite maintenance of treatment with cyclophosphamide. In other studies, complete or partial regression of KS has been observed following dosage reduction or cessation of corticosteroid therapy (Gange and Jones 1978; Leung et al. 1981; Scaparro et al. 1984).

To date, the study of Klepp et al. (1978) provides the only estimate of the potential risk of KS in association with immunosuppressive therapy. Data from the Norwegian cancer registry showed that, of 41 genuine cases of KS recorded over a 5-year period, six (15 per cent) had arisen in patients treated with corticosteroids and two of these had also received azathioprine. Gange and Jones (1978) reviewed the clinical features of patients developing KS during immunosuppressive therapy. They reported a marked reduction in the usual male–female ratio observed in spontaneous Kaposi's sarcoma, from 10:1 to around 2:1. The sites of KS nodules have been unremarkable, typically confined to the ankles and feet. Widespread lesions on all limbs and the trunk have been reported, however (Turnbull and Almeyda 1970; Klein et al. 1974), as have examples of extensive visceral involvement (Dantzig 1974; Gange and Jones 1978; Klepp et al. 1978).

The mechanisms by which corticosteroid therapy might predispose to the occurrence of KS remain speculative. Klepp et al. (1978) suggested that observations of spontaneous regression in KS point strongly to immunological mechanisms in its natural history, and that those mechanisms might be weakened by concurrent immunosuppressive treatment. Such a hypothesis would be in keeping with the regression of KS observed in many cases where immunosuppressive therapy has been withdrawn. Klepp and co-workers have also suggested that angiogenic factors may be released as a result of the chronic immunological reactivity of many disease states although, if this were true, it might be expected that the use of prednisolone and other immunosuppressive drugs should suppress such a mechanism. Erban and Sokas (1988) consider the lymphocytotoxic action of corticosteroids to be of major importance in this context.

Carcinoma of the uterine cervix

A possible association between SLE and the occurrence of cervical neoplasia has been suggested in several reports. Out of eight malignancies in 70 patients with SLE, reported by Canoso and Cohen (1974), five were cervical tumours, including three carcinomas *in situ* and two invasive lesions. Only one patient had received azathioprine and two had received no systemic therapy at all. Although this incidence of cervical tumours represented a 4-fold increase, it was not statistically significant. Lewis *et al.* (1976) reported six cervical cancers in 484 patients with SLE. Details of the tumours were not given but, when compared with four cervical cancers in 2867 patients with rheumatoid arthritis, this amounted to a 9-fold increase.

The impact of immunosuppressive therapy in the occurrence of cervical cancer was examined by Nyberg *et al.* (1981) who reported epithelial abnormalities varying from slight cellular atypia to invasive carcinoma in 19 of 80 female patients with SLE. This was significantly higher than nine similar cases which occurred in a control group of 80 patients treated without immunosuppression. The increased incidence was chiefly observed among 18 patients whose treatment was predominantly with azathioprine. Nine (50 per cent) of these were found to have varying degrees of cervical dysplasia.

While there is, thus, some suggestion of an inherently increased incidence of cervical tumours in SLE, possibly enhanced by treatment with azathioprine, there is little to indicate a heightened risk in other benign disorders. In a study of 202 patients treated with azathioprine for rheumatoid arthritis, the results relating to the incidence of cervical neoplasia were negative, with two cervical tumours in the treated group, compared with one in the same number of control patients (Silman *et al.* 1988) and there is thus no firm evidence linking azathioprine with the occurrence of cervical cancer.

Malignancy and the skin

Skin carcinomas following cytotoxic chemotherapy In an Australian study, Marshall (1974) reported neoplastic or pre-neoplastic skin changes in nine of 300 non-transplant patients treated with prednisolone and other immunosuppressive drugs, primarily for renal disease. Of the nine affected patients, eight developed squamous cell carcinoma on a background of multiple solar keratoses or keratoacanthoma, while the remaining patient developed multiple keratoses without overt malignant change. Five had received either cyclophosphamide or azathioprine and most had a history of chronic sunlight exposure. It was postulated that malignant transformation had been accelerated in skin where sun-induced preneoplastic changes had already occurred. Kinlen (1985) also reported an excess of squamous cell tumours in patients treated with azathioprine and cyclophosphamide, but in other studies the majority of skin tumours have been basal cell carcinomas (Baltus *et al.* 1983; Baker *et al.* 1987; Silman *et al.* 1988). In one of these studies (Baker *et al.* 1987), nine skin carcinomas (seven basal cell carcinomas) were diagnosed in 119 rheumatoid patients treated with cyclophosphamide, as against none in the control patients.

Malignancy following methotrexate therapy for psoriasis Methotrexate has been used for severe psoriasis and psoriatic arthropathy for over 30 years (Edmundson and Guy 1958). Dose schedules vary from small daily doses, under 5 mg, to weekly treatments of up to 25 mg, and total doses received vary from between a few hundred milligrams to 10 grams. Treatment is administered orally and is generally well tolerated. Long-term side effects are well recognized, however, the most serious being hepatotoxicity (Dahl *et al.* 1972), and patients who receive prolonged treatment require regular monitoring of liver function. Interest has also focused on the potential carcinogenicity of this treatment although there is little evidence for this at present. For example, its use in gestational trophoblastic disease, where cure rates approach 100 per cent, has been associated with no increase in the incidence of other malignancies (Rustin *et al.* 1983).

In a retrospective and prospective study of 224 patients with psoriasis, most of whom had received over 5 g of methotrexate in total, Bailin *et al.* (1975) found no overall increase in cancer although a number of uncommon tumours were observed, including one case each of AML, non-Hodgkin's lymphoma and oesophageal cancer. The authors commented that these tumours would be extremely unlikely to have arisen by chance in a population of just over 200 patients. Three basal cell carcinomas and one squamous cell skin tumour were also observed but excluded from analysis, since it was argued that in a patient group subjected to regular and frequent dermatological examination, such tumours would be diagnosed with greater frequency than in the general population. In a US national survey of 1650 psoriatic patients treated with methotrexate, Bergstresser *et al.* (1976) observed a low incidence of malignancy. Systemic malignancies, including one case of AML, were diagnosed in four patients and three others developed cutaneous tumours. The incidence rates for these tumours were not significantly higher than in the general population.

Other cancers

Evidence that immunosuppressive or cytotoxic therapy predisposes to the occurrence of other tumours is generally lacking. Kinlen (1985) observed a 5-fold increase in hepatocellular carcinoma among rheumatoid patients treated with immunosuppressive drugs. Baker *et al.* (1987) recorded seven lung cancers in 119 patients treated with cyclophosphamide for rheumatoid arthritis, but also five lung cancers in an equal number of untreated patients. Lung cancer has also been described in association with azathioprine treatment. Speerstra *et al.* (1982) observed four lung cancers among six malignant tumours in 91 patients treated for rheumatoid arthritis, but all affected patients were heavy smokers. In two other cohorts of rheumatoid patients, one group of 81 patients treated with cyclophosphamide (Baltus *et al.* 1983), the other group of 202 patients with azathioprine (Silman *et al.* 1988), a higher than expected number of lung cancers was observed. No details of smoking habits were available for the earlier cohort, and although the overall incidence of non-haematological tumours was significantly higher than in the untreated group in the later study, there were no significant excesses at individual cancer sites compared to an untreated control group.

In some studies, there has been a suggestion of an association between azathioprine therapy and the occurrence of breast cancer. Lhermitte *et al.* (1984) reported 10 cancers (five breast tumours) in 131 patients treated with azathioprine for multiple sclerosis. By comparison, only four malignancies had been observed in a control population of similar size. Silman *et al.* (1988) observed six breast cancers in the azathioprine-treated group of rheumatoid patients versus four in the control group.

Conclusions

It is clear that there is a sharp increase in the incidence of certain malignancies as a result of the immunosuppression induced artificially in the context of organ transplantation. This is particularly firmly established with respect to skin cancer, non-Hodgkin's lymphoma and Kaposi's sarcoma. Although these are quite distinct tumours, all of them may arise on a background of disordered immunity. The skin is exposed to constant antigenic challenge and functions as a major immunological organ in its own right. Impairment of the ability of the skin to respond appropriately to antigens may thus predispose to neoplastic change. The lymphoreticular system subserves many complex immunological interactions, and major disruption of these functions could well lead to lymphomatous transformation. While the association of Kaposi's sarcoma with the immune system is less certain, immunological factors nonetheless appear to predominate, as evidenced both by a tendency to spontaneous regression and the frequently observed regression of disease following cessation of immunosuppressive therapy.

To what extent does the state of organ transplantation *itself* expose the recipient to an increased risk of lymphoma? Possible mechanisms for the development of lymphoma have been reviewed above, and it is quite probable that a combination of factors is responsible. Penn (1986), however, suggests that the chronic antigenic stimulation by the allograft is important in the induction of lymphoid proliferation. In addition, there is evidence that the incidence of cerebral lymphoma is enhanced still further in patients who have received multiple organ transplants (Barnett and Schwartz 1974; Dorreen *et al.* 1988). On the other hand, since all such patients receive prolonged immunosuppressive therapy, it would be difficult to identify the extent to which transplantation and immunosuppressive therapy contribute separately to the risk of lymphoma. Nonetheless, as indicated above, the incidence of lymphoma appears to be greater among recipients of hepatic and cardiac allografts, for whom more profound immunosuppression has been required in order to prevent graft rejection.

Similar considerations may apply in other conditions, although to a lesser extent. The most consistent evidence of an increased incidence of malignant lymphoma has been demonstrated in rheumatoid arthritis and the related Sjögren's syndrome. For other inflammatory connective tissue diseases the evidence is patchy, and does not provide a well-established case. However, it is probable that the risk of lymphoma is further potentiated by the use of immunosuppressive and cytotoxic drugs. This has been most convincingly demonstrated by Kinlen and co-workers (Kinlen *et al.* 1979; Kinlen 1985) and applies to a wide range of inflammatory disorders.

The increased frequency of other, non-lymphoid tumours, can be more or less directly attributed to prior therapy. As in organ transplantation, this applies particularly to the occurrence of skin cancer and Kaposi's sarcoma and, to a lesser extent, to bladder cancer and acute myeloid leukaemia. The relationship between treatment and the development of other malignancies is not well-defined.

Clearly, the risk of treatment-induced malignancy needs to be set in the context in which potentially carcinogenic treatment is given. Overall, the problem is relatively small in magnitude but it has the most serious consequences among organ transplant recipients, for whom long-term immunosuppression remains at present an inescapable necessity. The use of cyclosporin A has permitted considerable reduction or even elimination of other immunosuppressive agents, however, and it may be that eventually a fall in the incidence of transplant-related lymphoma will be a consequence of this. In other conditions, it can be argued that there are sound indications for the use of immunosuppressive or cytotoxic drugs in severe or life-threatening inflammatory disease where few alternative therapeutic options remain. In all other circumstances, caution in the use of cytotoxic therapy should be advised. The potential risks of such treatment should be carefully weighed against a possibly unproven benefit, particularly in diseases where the place of immunosuppressive drugs has not been firmly established. In these cases, the most prudent counsel would be to adopt the simplest and least toxic treatment required to induce remission of symptoms and disease activity.

References

Alloub, M., Barr, B., Smart, G. E., Smith, I., McLaren, K., and Bunney, M. (1988). Risk of human papillomavirus infection (HPV) and cervical intraepithelial neoplasia (CIN) in a group of renal allograft recipients and characterisation of the types of HPV contained in their cervices. *Nephrology, Dialysis and Transplantation*, 3, 704.

Anderson, L. G. and Talal, N. (1971). The spectrum of benign to malignant lymphoproliferation in Sjögren's Syndrome. *Clinical and Experimental Immunology*, 9, 199–222.

Andreev, V. C. and Zlatkov, N. B. (1968). Systemic lupus erythematosus and neoplasia of the lymphoreticular system. *British Journal of Dermatology*, 80, 503–8.

Ansell, I. D. and Castro, J. E. (1975). Carcinoma of the bladder complicating cyclophosphamide therapy. *British Journal of Urology*, 47, 413–18.

Ansher, A. F., Melton, J. W., and Sliwinski, A. J. (1983). Bladder malignancy in a patient receiving low dose cyclophosphamide for treatment of rheumatoid arthritis. *Arthritis and Rheumatism*, 26, 804–5.

Bailin, P. L., Tindall, J. P., Roenigk, H. H., and Hogan, M. D. (1975). Is methotrexate therapy for psoriasis carcinogenic? A modified retrospective-prospective analysis. *Journal of the American Medical Association*, 232, 359–62.

Baker, G. L., Kahl, L. E., Zee, B. C., Stolzer, B. L., Agarwal, A., and Medsger, T. A. (1987). Malignancy following treatment of rheumatoid arthritis with cyclophosphamide. *American Journal of Medicine*, 83, 1–9.

Baltus, J. A. M., Boersma, J. W., Hartman, A. P., and Vandenbroucke, J. P. (1983). The occurrence of malignancies in patients with rheumatoid arthritis treated with cyclophosphamide: a controlled retrospective follow-up. *Annals of the Rheumatic Diseases*, 42, 368–73.

Barnett, L. B. and Schwartz, E. (1974). Cerebral reticulum cell sarcoma after multiple renal transplants. *Journal of Neurology, Neurosurgery and Psychiatry*, 37, 966–70.

Bergstresser, P. R., Schreiber, S. H., and Weinstein, G. D. (1976). Systemic chemotherapy for psoriasis. A national survey. *Archives of Dermatology*, 112, 977–81.

Beveridge, T. (1983). Cyclosporin A: an evaluation of clinical results. *Transplantation Proceedings*, 15, 433–7.

Bieber, C. P., et al. (1981). Complications in long term survivors of cardiac transplantation. *Transplantation Proceedings*, 13, 207–11.

Birkeland, S. A. (1983). Malignant tumors in renal transplant patients. The Scandia transplant material. *Cancer*, 51, 1571–5.

Boyle, D. J., Day, J. F., Kassan, S. S., Thomas, M. R., Robinson, W. A., and Steigerwald, J. C. (1981). Incidence of malignancy 10 years following cyclophosphamide use for rheumatoid arthritis. *Arthritis and Rheumatism*, 24 (Suppl.), 71.

Brunner, F. P., et al. (1989). Combined report on regular dialysis and transplantation in Europe XVIII 1987. *Nephrology, Dialysis and Transplantation*, 4 (Suppl. 2), 5–32.

Burnet, F. M. (1971). Immunological surveillance in neoplasia. *Transplantation Reviews*, 7, 3–20.

Cadman, E. C., Capizzi, R. L., and Bertino, J. R. (1977). Acute nonlymphocytic leukaemia. A delayed complication of Hodgkin's disease therapy: analysis of 109 cases. *Cancer*, **40**, 1280–96.
Calabro, J. J. (1967). Cancer and arthritis. *Arthritis and Rheumatism*, **10**, 553–67.
Cammarata, R. J., Rodnan, G. P., and Jenson, W. N. (1963). Systemic rheumatic disease and malignant lymphoma. *Archives of Internal Medicine*, **111**, 330–7.
Canoso, J. J. and Cohen, A. S. (1974). Malignancy in a series of 70 patients with systemic lupus erythematosus. *Arthritis and Rheumatism*, **17**, 383–90.
Chang, R. S., Lewis, J. P., Reynolds, R. D., Sullivan, M. J., and Neuman, J. (1978). Oropharyngeal excretion of Epstein–Barr virus by patients with lymphoproliferative disorders and by recipients of renal homografts. *Annals of Internal Medicine*, **88**, 34–40.
Chaplin, H. (1982). Lymphoma in primary chronic cold haemagglutinin disease treated with chlorambucil. *Archives of Internal Medicine*, **142**, 2119–23.
Cheeseman, S. H., et al. (1980). Epstein–Barr virus infections in renal transplant recipients. *Annals of Internal Medicine*, **93**, 39–42.
Cohen, M. G., Janssen, B., and Webb, J. (1988). Response of rheumatoid arthritis to chemotherapy for Hodgkin's disease in a patient with IgA deficiency and overlap connective tissue disease. *Annals of the Rheumatic Diseases*, **47**, 603–5.
Cox, P. J. (1979a). Cyclophosphamide cystitis: identification of acrolein as the causative agent. *Biochemical Pharmacology*, **28**, 2045–9.
Cox, P. J. (1979b). Cyclophosphamide cystitis and bladder cancer. A hypothesis. *European Journal of Cancer*, **15**, 1071–2.
Crawford, P. H., Edward, J. M. B., Sweny, P., Hoffbrand, A. V., and Janossy, G. (1981). Studies on long-term cell mediated immunity to Epstein–Barr virus in immunosuppressed renal allograft recipients. *International Journal of Cancer*, **28**, 705–9.
Cudworth, A. G. and Ellis, A. (1972). Malignant lymphoma and acute S.L.E. *British Medical Journal*, **3**, 291–2 (letter).
Dahl, M. G., Gregory, M. M., and Scheuer, P. J. (1972). Methotrexate hepatotoxicity in psoriasis: comparison of different dose regimens. *British Medical Journal*, **1**, 654–6.
Dale, G. A. and Smith, R. B. (1974). Transitional cell carcinoma of the bladder associated with cyclophosphamide. *Journal of Urology*, **112**, 603–4.
Dantzig, P. I. (1974). Kaposi sarcoma and polymyositis. *Archives of Dermatology*, **110**, 605–7.
De Bock, R. F. K. and Peetermans, M. E. (1977). Leukaemia after prolonged use of melphalan for non-malignant disease. *Lancet*, **i**, 1208–9.
Doll, R. and Kinlen, L. (1970) Immunosurveillance and cancer: epidemiological evidence. *British Medical Journal*, **4**, 420–2.
Dorreen, M. S., Ironside, J. W., Bradshaw, J. D., Jakubowski, J., Timperley, W. R., and Hancock, B. W. (1988). Primary intracerebral lymphoma: a clinicopathological analysis of 14 patients presenting over a 10 year period in Sheffield. *Quarterly Journal of Medicine*, **67**, 387–404.
Edmundson, W. F. and Guy, W. B. (1958). Treatment of psoriasis with folic acid antagonists. *Archives of Dermatology*, **78**, 200–3.
Elkon, K. B., et al. (1979). Hairy-cell leukaemia with polyarteritis nodosa. *Lancet*, **ii**, 280–2.

Elliot, R. W., Essenhigh, D. M., and Morley, A. R. (1982). Cyclophosphamide treatment of systemic lupus erythematosus: risk of bladder cancer exceeds benefit. *British Medical Journal*, **284**, 1160–1.

Erban, S. B. and Sokas, R. K. (1988). Kaposi's sarcoma in an elderly man with Wegener's granulomatosis treated with cyclophosphamide and corticosteroids. *Archives of Internal Medicine*, **148**, 1201–3.

Fairchild, W. V., Spence, C. R., Solomon, H. D., and Gangai, M. P. (1979). The incidence of bladder cancer after cyclophosphamide therapy. *Journal of Urology*, **122**, 163–4.

Farber, S. J., Sheon, R. P., Kirsner, A. B., and Finkel, R. I. (1979). Incidence of malignant disease in patients receiving cytotoxic therapy for rheumatoid arthritis. *Arthritis and Rheumatism*, **22**, 608 (abstract).

Forni, A. M., Koss, L. G., and Geller, W. (1964). Cytological study of the effect of cyclophosphamide on the epithelium of the urinary bladder in man. *Cancer*, **17**, 1348–55.

Fosdick, W. M., Parson, J. L., and Hill, D. F. (1969). Long-term cyclophosphamide (CP) therapy in rheumatoid arthritis: a progress report on six years' experience. *Arthritis and Rheumatism*, **12**, 663 (abstract).

Frizzera, G., Rosai, J., Dehner, L. P., Spector, B. D., and Kersey, J. H. (1980). Lymphoreticular disorders in primary immunodeficiencies: new findings based on an up-to-date histologic classification of 35 cases. *Cancer*, **46**, 692–9.

Fuchs, E. F., Kay, R., Poole, R., Barry, J. M., and Pearse, H. D. (1981). Uroepithelial carcinoma in association with cyclophosphamide ingestion. *Journal of Urology*, **126**, 544–5.

Gange, R. W. and Jones, E. W. (1978). Kaposi's sarcoma and immunosuppressive therapy: an appraisal. *Clinical and Experimental Dermatology*, **3**, 135–46.

Garvin, D. D. and Ball, T. P. (1981). Bladder malignancy in patient receiving cyclophosphamide for benign disease. *Urology*, **18**, 80–1.

Gibbons, R. B. and Westerman, E. (1988). Acute nonlymphocytic leukaemia following short-term, intermittent, intravenous cyclophosphamide treatment of lupus nephritis. *Arthritis and Rheumatism*, **31**, 1552–4.

Gill, P. S., et al. (1985). Primary central nervous system lymphoma in homosexual men. Clinical, immunologic and pathologic features. *American Journal of Medicine*, **78**, 742–8.

Goedert, J. J., Neefe, J. R., Smith, F. S., Stahl, N. I., Jaffe, E. S., and Fauci, A. S. (1981). Polyarteritis nodosa, hairy cell leukaemia and splenosis. *American Journal of Medicine*, **71**, 323–6.

Green, J. A., Dawson, A. A., and Walker, W. (1978). Systemic lupus erythematosus and lymphoma. *Lancet*, **ii**, 753–6.

Grünwald, H. W. and Rosner, F. (1979). Acute leukaemia and immunosuppressive drug use. A review of patients undergoing immunosuppressive therapy for non-neoplastic diseases. *Archives of Internal Medicine*, **139**, 461–6.

Hancock, B. W. and Ward, A. M. (ed.) (1985). *Immunological aspects of cancer*. Martinus Nijhoff, Boston.

Hansen, S. E. (1983). Carcinoma of the bladder in a patient treated with cyclophosphamide for rheumatoid arthritis. *Scandinavian Journal of Rheumatology*, **12**, 73–4.

Hanto, D. W., et al. (1983). Epstein–Barr virus induced polyclonal and monoclonal B-cell lymphoproliferative diseases occurring after renal transplantation: clinical, pathologic and virologic findings and implications for therapy. *Annals of Surgery*, **198**, 356–69.

Hanto, D. W., Frizzera, G., Gajl-Peczalska, K. J., and Simmons, R. L. (1985). Epstein–Barr virus, immunodeficiency and B-cell lymphoproliferation. *Transplantation*, **39**, 461–72.

Hardie, I. R., Strong, R. W., Hartley, L. C. J., Woodruff, P. W. H., and Clunie, G. J. A. (1980). Skin cancer in Caucasian renal allograft recipients living in a sub-tropical climate. *Surgery*, **87**, 177–83.

Harwood, A. R., *et al.* (1979). Kaposi's sarcoma in recipients of renal transplant. *American Journal of Medicine*, **67**, 759–65.

Hatchberg, F. H., Miller, G., and Schooley, R. T. (1983). Central nervous system lymphoma related to Epstein–Barr virus. *New England Journal of Medicine*, **309**, 745–8.

Hazleman, B. T. and De Silva, M. (1982). The comparative incidence of malignant disease in rheumatoid arthritis exposed to different treatment regimens. *Annals of the Rheumatic Diseases*, **41** (Suppl.), 12–17.

Heidelberger, K. P. and LeGolvan, D. P. (1974). Wiskott–Aldrich syndrome and cerebral neoplasia: report of a case with localized reticulum cell sarcoma. *Cancer*, **33**, 280–4.

Helweg-Larsen, S., Jakobsen, J., Boesen, F., and Arlien-Soborg, P. (1986). Neurological complications and concomitants of AIDS. *Acta Neurologica Scandinavica*, **74**, 467–74.

Hochberg, M. C. and Schulman, L. E. (1978). Acute leukaemia following cyclophosphamide therapy for Sjögren's syndrome. *Johns Hopkins Medical Journal*, **142**, 211–14.

Hoover, R. and Fraumeni, J. F. (1973). Risk of cancer in renal transplant recipients. *Lancet*, **ii**, 55–7.

Hoshaw, R. A. and Schwartz, R. A. (1980). Kaposi's sarcoma after immunosuppressive therapy with prednisone. *Archives of Dermatology*, **116**, 1280–2.

IARC (International Agency for Research on Cancer) (1990). *IARC Monographs on the evaluation of carcinogenic risks to humans*, Vol. 50, *Pharmaceutical drugs*. IARC, Lyon.

Isomaki, H. A., Hakulinen, T., and Joutsenlahti, U. (1982). Excess risk of lymphomas, leukaemia and myeloma in patients with rheumatoid arthritis. *Annals of the Rheumatic Diseases*, **41** (Suppl.), 34–6.

Jack, A. S., Boyce, B. F., and Lee, F. D. (1986). Malignant histiocytosis complicating rheumatoid arthritis: report of four cases. *Journal of Clinical Pathology*, **39**, 16–21.

Kahn, M. F., Arlet J., Bloch-Michel, H., Caroit, M., Chaouat, Y., and Renier, J. C. (1979). Leucémies aiguës après traitement par agents cytotoxiques en rhumatologie. 19 observations chez 2006 patients. *Nouvelle Presse Médicale*, **8**, 1393–7.

Kapadia, S. B. and Kaplan, S. S. (1978). Acute myelogenous leukaemia following immunosuppressive therapy for rheumatoid arthritis. *American Journal of Clinical Pathology*, **70**, 301–2.

Kassan, S. S., *et al.* (1978). Increased risk of lymphoma in sicca syndrome. *Annals of Internal Medicine*, **89**, 888–92.

Kelly, G. E., Mahoney, J. F., Sheil, A. G. R., Meikle, W. D., Tiller, D. S., and Horvath, J. (1987). Risk factors for skin carcinogenesis in immunosuppressed kidney transplantation recipients. *Clinical Transplantation*, **1**, 271–7.

Kinlen, L. J. (1985). Incidence of cancer in rheumatoid arthritis and other disorders after immunosuppressive treatment. *American Journal of Medicine*, **78** (Suppl. 1A), 44–9.

Kinlen, L. J., Sheil, A. G. R., Peto, J., and Doll, R. (1979). Collaborative United Kingdom/Australasian study of cancer in patients treated with immunosuppressive drugs. *British Medical Journal*, **2**, 1461–6.

Kinlen, L. J. Peto, J., Doll, R., and Sheil, A. G. R. (1981). Cancer in patients treated with immunosuppressive drugs. *British Medical Journal*, **282**, 474.

Kinlen, L., Doll, R., and Peto, J. (1983). The incidence of tumours in human transplant recipients. *Transplantation Proceedings*, **15**, 1039–42.

Kirsner, A. B., Farber, S. J., Sheon, R. P., and Finkel, R. I. (1982). The incidence of malignant disease in patients receiving cytotoxic therapy for rheumatoid arthritis. *Annals of the Rheumatic Diseases*, **41** (Suppl.), 32–3.

Klein, G. and Purtilo, D. (1981). Summary: symposium on Epstein–Barr virus-induced lymphoproliferative disorders in immunodeficient patients. *Cancer Research*, **41**, 4302–4.

Klein, M. B., Pereira, F. A., and Kantor, I. (1974). Kaposi's sarcoma complicating systemic lupus erythematosus treated with immunosuppression. *Archives of Dermatology*, **110**, 602–4.

Klepp, O., Dahl, O., and Stenwig, J. T. (1978). Association of Kaposi's sarcoma and prior immunosuppressive therapy. A 5-year material of Kaposi's sarcoma in Norway. *Cancer*, **42**, 2626–30.

Lea, A. J. (1964). An association between the rheumatic diseases and the reticuloses. *Annals of the Rheumatic Diseases*, **23**, 480–4.

Lebranchu, Y., et al. (1980). Acute monoblastic leukaemia in child receiving chlorambucil for juvenile rheumatoid arthritis. *Lancet*, **i**, 649.

Leung, F., Fam, A. G., and Osoba, D. (1981). Kaposi's sarcoma complicating corticosteroid therapy for temporal arteritis. *American Journal of Medicine*, **71**, 320–2.

Lewis, R. B., Castor, C. W., Knisley, R. E., and Bole, G. G. (1976). Frequency of neoplasia in systemic lupus erythematosus and rheumatoid arthritis. *Arthritis and Rheumatism*, **19**, 1256–60.

Lewis, P., Hazleman, B. L., Hanka, R., and Roberts, S. (1980). Cause of death in patients with rheumatoid arthritis with particular reference to azathioprine. *Annals of the Rheumatic Diseases*, **39**, 457–61.

Lhermitte, F., Marteau, R., and Roullet, E. (1984). Not so benign long-term immunosuppression in multiple sclerosis. *Lancet*, **i**, 276–7 (letter).

Lipinski, M., Turz, T., Kresi, H., Finale, Y., and Amiel, J.-L. (1980). Dissociation of natural killer cell activity and antibody dependent cell-mediated cytotoxicity in kidney allograft recipients receiving high dose immunosuppressive therapy. *Transplantation*, **29**, 214–18.

Lipsmeyer, E. A. (1972). Development of malignant cerebral lymphoma in a patient with systemic lupus erythematosus treated with immunosuppression. *Arthritis and Rheumatism*, **15**, 183–6.

Love, R. R. and Sowa, J. M. (1975). Myelomonocytic leukaemia following cyclo-

phosphamide therapy of rheumatoid disease. *Annals of the Rheumatic Diseases*, **34**, 534–5.
MacLeod, A. M. and Catto, G. R. D. (1988). Cancer after transplantation. *British Medical Journal*, **297**, 4–5.
Marshall, V. (1974). Premalignant and malignant skin tumours in immunosuppressed states. *Transplantation*, **17**, 272–5.
McGeown, M. G., *et al.* (1988). Ten-year results of renal transplantation with azathioprine and prednisolone as only immunosuppression. *Lancet*, **i**, 983–5.
Miller, D. G. (1967). The association of immune disease and malignant lymphoma. *Annals of Internal Medicine*, **66**, 507–21.
Motulsky, A. G., Weinberg, S., Saphir, O., and Rosenberg, E. (1952). Lymph nodes in rheumatoid arthritis. *Archives of Internal Medicine*, **90**, 660–7.
Nyberg, G., Eriksson, O., and Westberg, N. G. (1981). Increased incidence of cervical atypia in women with systemic lupus erythematosus treated with chemotherapy. *Arthritis and Rheumatism*, **24**, 648–50.
Patchell, R. A. (1988). Primary central nervous system lymphoma in the transplant patient. *Neurologic Clinics*, **6**, 297–303.
Pedersen-Bjergaard, J., *et al.* (1987). Risk of therapy-related leukaemia and preleukaemia after Hodgkin's disease. *Lancet*, **ii**, 83–8.
Penn, I. (1981). Depressed immunity and development of cancer. *Clinical and Experimental Immunology*, **46**, 459–74.
Penn, I. (1982). Malignancies following the use of cyclosporin A in man. *Cancer Surveys*, **1**, 621–4.
Penn, I. (1983*a*). Renal transplantation in patients with pre-existing malignancies. *Transplantation Proceedings*, **15**, 1079–82.
Penn, I. (1983*b*). Kaposi's sarcoma in immunocompromised patients. *Journal of Clinical and Laboratory Immunology*, **12**, 1–10.
Penn, I. (1986). The occurrence of malignant tumours in immunosuppressed states. *Progress in Allergy*, **37**, 259–300.
Penn, I. (1988*a*). Cancers after cyclosporin therapy. *Transplantation Proceedings*, **20** (Suppl. 1), 276–9.
Penn, I. (1988*b*). Secondary neoplasms as a consequence of transplantation and cancer therapy. *Cancer Detection and Prevention*, **12**, 39–57.
Penn, I., Hammond W., Brettschneider, L., and Starzl, T. E. (1969). Malignant lymphomas in transplantation patients. *Transplantation Proceedings*, **1**, 106–11.
Pitt, P. I., Sultan, A. H., Malone, M., Andrews, V., and Hamilton, E. B. D. (1987). Association between azathioprine therapy and lymphoma in rheumatoid disease. *Royal Society of Medicine Journal*, **80**, 428–9.
Plotz, P. H., *et al.* (1979). Bladder complications in patients receiving cyclophosphamide for systemic lupus erythematosus or rheumatoid arthritis. *Annals of Internal Medicine*, **91**, 221–3.
Prior, P. (1985). Cancer and rheumatoid arthritis: epidemiologic considerations. *American Journal of Medicine*, **78** (Suppl. 1A), 15–21.
Prior, P., Symmons, D. P. M., Hawkins, C. F., Scott, D. L., and Brown, R. (1984). Cancer morbidity in rheumatoid arthritis. *Annals of the Rheumatic Diseases*, **43**, 128–31.
Richtsmeier, A. J. (1975). Urinary-bladder tumors after cyclophosphamide. *New England Journal of Medicine*, **293**, 1045–6.

Rieche, K. (1984). Carcinogenicity of antineoplastic agents in man. *Cancer Treatment Reviews*, **11**, 39–67.
Roberts, M. M. and Bell, R. (1976). Acute leukaemia after immunosuppressive therapy. *Lancet*, **ii**, 768–70.
Rowley, J. D., Golomb, H. M., and Vardiman, J. W. (1981). Nonrandom chromosome abnormalities in acute leukaemia and dysmyelopoietic syndromes in patients with previously treated malignant disease. *Blood*, **58**, 759–67.
Rustin, G. J. S., Rustin, F., Dent, J., Booth, M., Salt, S., and Bagshawe, K. (1983). No increase in second tumours after cytotoxic chemotherapy for gestational trophoblastic tumours. *New England Journal of Medicine*, **308**, 473–6.
Scaparro, E., Borghi, S., and Rebora, A. (1984). Kaposi's sarcoma after immunosuppressive therapy for bullous pemphigoid. *Dermatologica*, **169**, 156–9.
Scharf, J., Nahir, M., Eidelman, S., Jacobs, R., and Levin, D. (1977). Carcinoma of the bladder with azathioprine therapy. *Journal of the American Medical Association*, **237**, 152.
Schiff, H. I., Finkel, M., and Schapria, H. E. (1982). Transitional cell carcinoma of the ureter associated with cyclophosphamide therapy for benign disease. *Journal of Urology*, **128**, 1023–4.
Schmähl, D. (1977). Carcinogenic action of anticancer drugs with special reference to immunosuppression. *Cancer*, **40**, 1927–9.
Schmähl, D., Habs, M., Lorenz, M., and Wagner, I. (1982). Occurrence of second tumours in man after anticancer drug treatment. *Cancer Treatment Reviews*, **9**, 167–94.
Schottstaedt, M. W., Hurd, E. R., and Stone, M. J. (1987). Kaposi's sarcoma in rheumatoid arthritis. *American Journal of Medicine*, **82**, 1021–6.
Seidenfeld, A. M., Smythe, H. A., Ogryzlo, M. A., Urowitz, M. B., and Dotten, D. A. (1976). Acute leukaemia in rheumatoid arthritis treated with cytotoxic agents. *Journal of Rheumatology*, **3**, 295–304.
Seltzer, S. E., Benazzi, R. B., and Kearney, G. P. (1978). Cyclophosphamide and carcinoma of the bladder. *Urology*, **11**, 352–6.
Sharpstone, P., Ogg, C. S., and Cameron, J. S. (1969). Nephrotic syndrome due to primary renal disease in adults: II. A controlled trial of prednisolone and azathioprine. *British Medical Journal*, **2**, 535–9.
Sheibani, K., Bukowski, R. M., Tubbs, R. R., Savage, R. A., Sebek, B. A., and Hoffman, G. C. (1980). Acute nonlymphoctyic leukaemia in patients receiving chemotherapy for non-malignant diseases. *Human Pathology*, **11**, 175–9.
Sheil, A. G. R. (1986). Cancer after transplantation. *World Journal of Surgery*, **10**, 389–96.
Silman, A. J., Petrie, J., Hazleman, B., and Evans, S. J. W. (1988). Lymphoproliferative cancer and other malignancy in patients with rheumatoid arthritis treated with azathioprine: a 20-year follow up study. *Annals of the Rheumatic Diseases*, **47**, 988–92.
Silvergleid, A. J. and Schrier, S. L. (1974). Acute myelogenous leukaemia in two patients treated with azathioprine for non-malignant disease. *American Journal of Medicine*, **57**, 885–8.
Snider, W. D., Simpson, D. M., Aronyk, K. E., and Nielsen, S. L. (1983). Primary lymphoma of the nervous system associated with Acquired Immune-Deficiency Syndrome. *New England Journal of Medicine*, **308**, 45 (letter).

Snyder, R. A. and Schwartz, R. A. (1982). Telangiectatic Kaposi's sarcoma. *Archives of Dermatology*, **118**, 1020–1.
Speerstra, F., Boerbooms, A. M. T. H., Van de Putte, L. B. A., Van Beusekom, H. J., Kruijsen, M. W. M., and Vandenbroucke, J. P. (1982). Side effects of azathioprine treatment in rheumatoid arthritis, *Annals of the Rheumatic Diseases*, **41** (Suppl.), 37–9.
Startzl, T. E., et al. (1984). Reversibility of lymphomas and lymphoproliferative lesions developing under cyclosporin-steroid therapy. *Lancet*, **i**, 583–7.
Stillwell, T. J., Benson, R. C., DeRemee, R. A., McDonald, T. J., and Weiland, L. H. (1988). Cyclophosphamide-induced bladder toxicity in Wegener's granulomatosis. *Arthritis and Rheumatism*, **31**, 465–70.
Symmons, D. P. M. (1985). Neoplasms of the immune system in rheumatoid arthritis. *American Journal of Medicine*, **78** (Suppl. 1A), 22–8.
Symmons, D. P. M., et al. (1984). Lymphoproliferative malignancy in rheumatoid arthritis: a study of 20 cases. *Annals of the Rheumatic Diseases*, **43**, 132–5.
Talal, N. and Bunim, J. J. (1964). The development of malignant lymphoma in the course of Sjögren's syndrome. *American Journal of Medicine*, **36**, 529–40.
Talal, N., Sokoloff, L., and Barth, W. F. (1967). Extrasalivary lymphoid abnormalities in Sjögren's syndrome (reticulum cell sarcoma, "Pseudolymphoma", macroglobulinaemia). *American Journal of Medicine*, **43**, 50–65.
Turnbull, A. and Almeyda, J. (1970). Idiopathic thrombocytopenic purpura and Kaposi's sarcoma. *Royal Society of Medicine Journal*, **63**, 603–5.
Tye, M. J. (1970). Bullous pemphigoid and Kaposi's sarcoma. *Archives of Dermatology*, **101**, 690–1.
Urowitz, M. B., Smythe, H. A., Able, T., Norman, C. S., and Travis, C. (1981). Long-term effects of azathioprine in rheumatoid arthritis. *Annals of the Rheumatic Diseases*, **41** (Suppl.), 18–22.
Vaughan, J. H. (1985). Immune system in rheumatoid arthritis: possible implications in neoplasms. *American Journal of Medicine*, **78** (Suppl. 1A), 6–11.
Wall, R. L. and Clausen, K. P. (1975). Carcinoma of the urinary bladder in patients receiving cyclophosphamide. *New England Journal of Medicine*, **293**, 271–3.
Weiner, S. R., Noritake, D. T., and Paulus, H. E. (1988). Kaposi's sarcoma in a patient with rheumatoid arthritis and polymyositis treated with corticosteroids. *Western Journal of Medicine*, **148**, 702–3.
Weiss, V. C. and Serushan, M. (1982). Kaposi's sarcoma in a patient with dermatomyositis receiving immunosuppressive therapy. *Archives of Dermatology*, **118**, 183–5.
Wessel, G., Abendroth, K., and Wisheit, M. (1988). Malignant transformation during immunosuppressive therapy (azathioprine) of rheumatoid arthritis and systemic lupus erythematosus. A retrospective study. *Scandinavian Journal of Rheumatology*, **67** (Suppl.), 73–5.
Whaley, K., Webb, J., McAvoy, B. A., Hughes, G. R. V., Lee, P., MacSweeney, R. N. M., and Buchanan, W. W. (1973). Sjøgren's syndrome. 2. Clinical associations and immunological phenomena. *Quarterly Journal of Medicine*, **167**, 513–48.
Worth, P. H. L. (1971). Cyclophosphamide and the bladder. *British Medical Journal*, **3**, 182 (letter).

Young, C. L., Adamson, T. C., Vaughan, J. H., and Fox, R. I. (1984). Immunohistologic characterisation of synovial membrane lymphocytes in rheumatoid arthritis. *Arthritis and Rheumatism*, **27**, 32–9.

Zulman, J., Jaffe, R., and Talal, N. (1978). Evidence that the malignant lymphoma of Sjögren's syndrome is a monoclonal B-cell neoplasm. *New England Journal of Medicine*, **299**, 1215–20.

6
Childhood malignancy

JILLIAN M. BIRCH

Introduction

Malignancy in childhood is rare, but it accounts for a disproportionate loss of years of life. In England and Wales for many years, neoplasms represented the most common natural cause of death in children aged 1–14 years, being exceeded only by accidents, but with improvements in survival in recent years neoplasms now rank third as a cause of childhood mortality, after accidents and congenital malformations. Approximately one in 650 children will develop a malignancy before their 15th birthday (Birch *et al.* 1988). Table 6.1 shows the incidence of the main types of childhood cancer. The types and distribution of tumours seen in children are very different from those in adults, and the more common types of childhood malignancy occur almost exclusively in children, or at least are very uncommon beyond 15 years of age. Furthermore, these specific childhood malignancies typically present themselves under the age of five (Fig. 6.1), and many examples are seen in the first year of life. Many paediatric tumours have histological features reminiscent of fetal development, and are termed 'embryonal'. The early onset of these tumours, together with their embryonal nature, suggest a prenatal origin, and genetic factors may play an important role in their development.

The International Agency for Research on Cancer has recently published the results of a coordinated worldwide study of the incidence of childhood cancer in more than 50 countries (Parkin *et al.* 1988*a*). This study shows that childhood cancer incidence rates in most populations fall in a fairly narrow range, between 100 and 150 cases per million children per year, but there is considerable variation in the incidence of individual tumour types between populations. For example, the rate for Hodgkin's disease (age-standardized to the world standard population for age groups under 15 years) was 0.7 in Osaka, Japan, and 10.8 in Costa Rica, while the corresponding figure for England and Wales was 4.1. Wilms' tumour has been proposed by Innis (1972) as an 'index tumour', i.e. one with a uniform incidence throughout the world, but the international study has shown a 3- to 4-fold difference in the incidence of Wilms' tumour between black populations and those of East Asia, with white populations being intermediate (Parkin *et al.* 1988*b*). These unexpectedly large international variations in the incidence of childhood tumours may indicate that environ-

Table 6.1 Incidence of cancer in children aged 0–14 years: Manchester Children's Tumour Registry, 1976–1986

Diagnosis	Rate per million person years
Leukaemias	
Acute lymphocytic leukaemia	26.9
Acute non-lymphocytic leukaemia	5.8
Other leukaemia	1.6
Lymphomas	
Hodgkin's disease	4.5
Non-Hodgkin's lymphoma	5.9
Other reticuloendothelial system tumours	3.4
Brain and spinal	
Ependymoma	3.3
Astrocytoma	9.8
Medulloblastoma	6.4
Other central nervous system tumours	8.0
Neuroblastoma	5.9
Retinoblastoma	2.1
Wilms' tumour	5.5
Hepatoblastoma	0.6
Bone	
Osteosarcoma	2.4
Ewing's tumour	1.8
Soft tissue sarcoma	5.7
Germ cell tumours	3.9
Carcinoma and melanoma	3.6
Other and unspecified tumours	2.2
All cancers	109.3

mental factors play a more prominent role in their aetiology than has been thought in the past, and interactions between genetic and environmental factors may occur.

Knudson (1976) has proposed a mutation model for the pathogenesis of retinoblastoma, which was subsequently extended to other childhood embryonal cancers. This model proposes that tumours arise as a result of two mutational events. After the first event, a mutated cell will be in an intermediate, precancerous state. The second mutational event then results in complete malignant transformation and eventual development of a tumour. Under this model, the first event in a hereditary retinoblastoma would occur in a parental germ cell, and the second event in a somatic cell of the child. In sporadic cases of retinoblastoma, however, both mutational

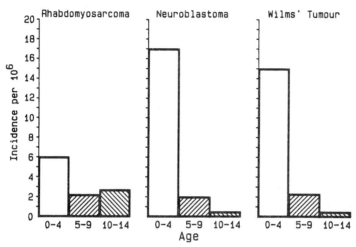

Fig. 6.1 Age-specific incidence of selected childhood embryonal tumours showing highest incidence in children aged less than five years.

events would occur in a somatic cell. The essential feature of this model is that both hereditary and sporadic forms of disease arise via the same genetic steps at the cellular level, but in the hereditary form of the disease all cells in a target tissue are in the premalignant state, and the probability of malignant change in one of these target cells is correspondingly high; this results in a clinically recognizable dominant pattern of inheritance.

With recent advances in DNA technology, it has been possible to demonstrate the validity of this hypothesis. In retinoblastoma, the responsible gene is located on chromosome 13, in band q14. The first event can be either deletion of the normal, dominant gene at this locus from one of the pair of chromosomes 13; or else a mutation of the gene to an abnormal, recessive form. The second event may be a mutuation or deletion of the second gene (at the homologous locus of the other chromosome), or else some recombinational event resulting in loss of the normal gene. It is implicit therefore that at the cellular level, such a gene is recessive, and it is the absence of the normal gene which is important for tumour development (Cavenee *et al.* 1983). Such genes are termed 'tumour suppressor genes' or anti-oncogenes. Exposure to mutagenic environmental agents, including radiation and certain drugs (see also Chapters 2–5, this volume), may thus influence the mutation rate, or else increase the likelihood of gene deletion or recombination at susceptible loci.

Relatively few large studies of aetiology in childhood cancer have been carried out, and these have identified few consistent risk factors. Among those which have been identified are constitutional chromosomal abnormalities including trisomy 21 (Down's syndrome) with acute leukaemia

(Robison et al. 1984); and the 11p deletion syndrome with Wilms' tumour (Riccardi et al. 1978). Certain congenital malformation syndromes are associated with a high risk of childhood cancer, for example Beckwith–Wiedemann syndrome, which predisposes to Wilms' tumour, adrenal cortical tumours, hepatoblastoma and other childhood malignancies (Sotelo-Avila et al. 1980).

Medical exposures to ionizing radiations, drugs, and certain other medical procedures are among the extrinsic factors which have been associated with a higher risk of childhood malignancy, and it is exposure to this group of risk factors which forms the subject of the present discussion. These exposures will be considered under three headings: first, prenatal or *in utero* exposures; second, postnatal exposures for benign conditions; and finally, treatments for childhood malignancy. In the present context 'medical treatment' is taken to mean any medical procedure which may influence carcinogenic risk.

Prenatal exposure

In utero *irradiation*

The association between prenatal exposure to diagnostic X-rays and childhood malignancy was first reported more than 30 years ago (Stewart et al. 1956), but in spite of numerous subsequent studies and analyses of the data, the question of whether the observed excess of cancers among children exposed to diagnostic obstetric X-rays is causally linked to X-ray exposure is still controversial.

Two large studies have explored the relationship between prenatal X-rays and childhood cancer: the Oxford Survey of Childhood Cancer (OSCC), carried out in the UK; and a study based in the US (MacMahon 1962). Several smaller studies have also been carried out. Each of the two large studies demonstrated an approximately 40 per cent increase in the risk of dying from malignancy before the tenth birthday in children who had been irradiated *in utero*. In some of the smaller studies, including the study by Court Brown et al. (1960) in which the leukaemia experience of 40 000 children irradiated *in utero* was observed in a prospective fashion, no excess risk was demonstrated, but sample sizes were such that an increased risk of the order observed in the two large studies could not be excluded. A weighted combination of all the studies published up to 1964 also demonstrated an excess risk of about 40 per cent. This increased risk applied both to leukaemias and to solid tumours (MacMahon and Hutchison 1964). Results of an extension to MacMahon's original study, based on approximately double the original study population, were broadly compatible with the original findings. Risk ratios (calculated over both phases of the study) were 1.5 for leukaemias and 1.3 for solid tumours (Monson and MacMahon 1984). Using the OSCC data, Stewart and Kneale (1970) demonstrated a dose-response relationship between X-irradiation and childhood cancer mortality, where 'dose' was based only on the number of exposures, since dose estimates were not directly available. They predicted 3–8 (mean 5.72) 'extra' cancer deaths per 100 person-Gy by the age of 10 from irradiation of the fetus.

The argument against the association between exposure to diagnostic X-rays *in utero* and an increased risk of death from childhood cancer is based on three considerations. First, there is no evidence of any excess of cancer deaths before the age of 10 among Japanese children exposed *in utero* to less than 5 Gy from the atomic bombs at Hiroshima and Nagasaki, whereas 5.2 extra cancer deaths would have been predicted from even the lower limit of Stewart and Kneale's model (Jablon and Kato 1970). A more recent analysis of the Japanese data, however, based on follow-up of 1630 individuals exposed *in utero* at the time of the atomic bombs, does show an excess of cancer, which is also related to dose, but the excess is entirely due to cancers diagnosed *over* the age of 10 years, and there were no childhood leukaemias. One case of cancer was found under 10 years of age: a liver tumour in a 6-year-old who was exposed to 1.4 Gy *in utero*. The 95 per cent upper confidence limit of excess cancer risk up to 10 years of age, based on the Japanese data, is 2 per 100 person Gy, compared to the lower limit of 3 per 100 person Gy predicted on the OSCC results (Jablon and Kato 1970; Yoshimoto *et al*. 1988). Thus, although excess cancers were found in this extended study of A-bomb survivors exposed *in utero*, and there was evidence of greater susceptibility than in those exposed postnatally, the cancers were mainly solid tumours in adults. These observations are somewhat at variance with what might have been predicted from the OSCC data.

The second argument results from the conflict between the OSCC results and animal experimentation data: this has been reviewed by the United Nations Scientific Committee on the Effects of Atomic Radiation (UNSCEAR 1977). Irradiation of mice and rats *in utero* appears to be no more carcinogenic than irradiation of adult animals.

Third, the idea of a causal association is challenged by Miller's (1969) observation: he pointed out that it was peculiar that *in utero* irradiation should increase the risk of leukaemias and solid malignancies to the same extent, when the relative risks for leukaemia were about three times higher than the risks for other cancers (about 24-fold compared to 8-fold) in children who were less than 10 years old when irradiated by the Hiroshima and Nagasaki bombs (Beebe *et al*. 1978).

It has been suggested that the association between cancer and X-irradiation of the fetus is simply due to confounding by a third factor, which is causally associated with increased cancer risk and which also results in an increased likelihood of maternal radiographic investigations (Miller 1969; Burch 1970, 1974). Children dying from cancer would thus come from a different subset of the population than the selected comparison group (Totter and MacPherson 1981).

Extensive analyses of both the OSCC and the US data, however, allowing for possible confounders, have failed to identify any such confounding factor.

Mole (1974) examined the association between *in utero* irradiation and cancer in twins. The radiography rate among singleton pregnancies in the OSCC series was 10 per cent, and 55 per cent among twin pregnancies. The reason for radiographic examination in a twin pregnancy in most cases is known, i.e. the twin pregnancy itself, thereby eliminating the suspected third factor. The excess of leukaemia and solid tumours in the X-rayed twins was similar to that in X-rayed singletons, and Mole concluded that this was strong evidence for irradiation as the cause of the excess. More recently, Harvey *et al.* (1985) reported a case-control study of the association in more than 32 000 twins. The twins were identified from a birth certificate search in the State of Connecticut between 1930 and 1969, and these were linked with notifications of cancer cases between 1935 and 1981. Twins with cancer were matched with four twin controls who had not developed cancer, and the *in utero* X-ray experience of the two groups was ascertained and compared. The results demonstrated a relative risk of 2.4 among those twins exposed to X-rays *in utero*, and the authors suggest that, although the study was based on small numbers, the results provide further evidence that low dose prenatal irradiation may increase the risk of childhood cancer.

MacMahon (1985) has commented that the analyses by Mole (1974) and Harvey *et al.* (1985) 'push us some way towards thinking of the association in causal terms, but by no means all the way', and raised three areas of uncertainty. The first is that of the small numbers involved, especially in the Connecticut study; the second is whether Mole's inference that the reason for irradiation in twin pregnancies is known is correct, and although multiple pregnancy would have been a strong indication for obstetric radiography, this does not exclude all other indications for radiographic procedures; while the third is the observation that, in general, the incidence of cancer appears to be lower in twins than in singletons, whereas if twins are irradiated more often than singletons, and X-ray exposure increases the risk, then twins would be expected to have a higher cancer incidence. MacMahon concludes that it seems unlikely that the issue will ever be resolved, and that the association between fetal irradiation and childhood cancer will remain 'the source of more confusion than enlightenment'.

Exposure to drugs in utero

The transplacental induction of tumours in experimental animals is well known, particularly neurogenic tumours induced transplacentally in rodents by nitrosoureas (Druckrey et al. 1966; Wechsler et al. 1979), and the possibility that a proportion of childhood malignancies are the result of exposure of the mother to carcinogenic substances during pregnancy has been raised. The main feature of the animal experiments is that the fetus appears to be very much more sensitive to carcinogenesis by these substances than the adult animal. In Druckrey's work, for example, the offspring of rats exposed to N-ethylnitrosourea during pregnancy were about 50 times more susceptible than adult animals, and they frequently died of multiple neurogenic tumours within the first year following a single prenatal exposure.

The first clear example of human transplacental carcinogenesis was described by Herbst et al. in 1971. These workers demonstrated a strong association between maternal exposure to large doses of diethylstilboestrol (DES) during pregnancy and the development of clear cell adenocarcinoma of the vagina in their daughters. Vaginal clear cell adenocarcinoma is rare, and was almost never seen in younger patients prior to 1965 (Herbst and Scully 1970). The synthetic hormone DES was given to large numbers of women in the 1940s and 1950s, particularly in the US, to prevent recurrent miscarriage. Although the therapeutic effectiveness of DES in that clinical situation was doubted as early as the 1950s, its use was continued by some physicians throughout the 1960s (Heinonen 1973).

Following the identification of the initial cluster of cases associated with intrauterine exposure to DES, a registry which collected clinical, epidemiologic, and pathologic details of cases of clear cell adenocarcinoma of the vagina and cervix was set up. Cases included in the register to June 1980 were reviewed by Herbst (1981). Maternal histories were available on 389 of the 429 cases included in the register, and among these 243 (63 per cent) had evidence of exposure to DES or related substances *in utero*. The incidence appears to have peaked in the early 1970s. The subsequent decrease in incidence may be due, at least in part, to the decline in the use of DES for obstetric purposes during the 1960s. The youngest woman to develop a clear cell adenocarcinoma after prenatal exposure to DES was 7 years old at the time of diagnosis, and the oldest was 30 years. The peak incidence is between 14 and 21 years of age, and the median age is 19 years. In addition to carcinomas, non-malignant epithelial changes have been observed in the genital tract of females exposed to DES *in utero*. Male offspring of women treated with DES in pregnancy do not appear to be at increased risk of malignancy (Herbst et al. 1979).

The risk of clear cell adenocarcinoma of the vagina or cervix developing in females prenatally exposed to DES is estimated to be of the order of 0.14 to 1.4 per 1000 up to the age of 24 years. Precise estimates of risk are not possible due to lack of data on the number of women treated with DES during pregnancy (Herbst et al. 1977). Although few data are available on cancer risks among the women themselves who were treated with DES during pregnancy, it seems that the excess risk of cancer, if it exists, is small (Bibbo et al. 1978; Brian et al. 1980). The experience from the DES episode suggests that the human fetus, like the animal fetus, is more susceptible to carcinogenesis than the adult.

A second candidate human transplacental carcinogen is diphenylhydantoin, a drug used to treat epilepsy. There are now a number of case reports of malignancies in the offspring of women taking diphenylhydantoin during pregnancy. In five cases the tumour was neuroblastoma or ganglioneuroblastoma; the remaining cases include mesenchymoma, melanotic ectodermal tumour, Wilms' tumour, and ependymoblastoma. These cases are reviewed by Lipson and Bale (1985). In six of the cases diphenylhydantoin was taken in combination with phenobarbitone or methylphenobarbitone. Six of the infants, including all those with neuroblastoma or ganglioneuroblastoma, also showed the fetal hydantoin syndrome. This syndrome was first described by Hanson and Smith (1975), and its features include prenatal and postnatal growth deficiency, motor and developmental delays, mental retardation, craniofacial anomalies, and distal phalangeal hyperplasia. Diphenylhydantoin exposure during pregnancy is independently associated with an increased incidence of other major birth defects, for example congenital heart disease, and cleft lip and palate (Majewski et al. 1981). Diphenylhydantoin thus appears to have both teratogenic and carcinogenic properties. It is of interest that in a number of the cases reviewed by Lipson and Bale (1985) the women had also taken phenobarbitone (or methylphenobarbitone, which is metabolized to phenobarbitone by the liver), since phenobarbitone has itself been associated with a 2-fold risk of brain tumours in transplacentally exposed children (Gold et al. 1978). The OSCC also showed an overall relative risk of 2.1 for all childhood cancer in children whose mothers took 'sedatives' during pregnancy (Kinnier Wilson et al. 1981). In a case-control study of childhood leukaemias and lymphomas, no association was found with barbiturates during pregnancy, but there was an excess of leukaemias among children whose mothers took barbiturates during labour (McKinney et al. 1987): this result was based on information abstracted from obstetric notes, and was not therefore subject to biased recall of drugs by the mother after the event. A case-control study of neuroblastoma conducted in the US found a significant excess of maternal use of neurally active drugs during pregnancy, including three case mothers but no control mothers who reported taking phenobarbitone at some time during their pregnancy (Kramer et al. 1987). In this study, however, maternal recall of drug exposures was not validated from medical records, although the authors considered that there was no evidence of recall bias.

In summary, the question of whether diphenylhydantoin and phenobarbitone possess transplacental carcinogenic properties, particularly for tumours of the peripheral and central nervous systems, is not resolved by the available evidence. This should be the subject of further study.

Other reported associations between maternal use of drugs during pregnancy and development of malignant disease in the offspring include narcotic analgesics with leukaemias (McKinney et al. 1987); diuretics and sex hormones with neuroblastoma (Kramer et al. 1987); diuretics, antihistamines, and general anaesthesia with brain tumours (Preston-Martin et al. 1982); and, more generally, use of any drug in pregnancy with the overall risk of childhood cancer (Kinnier-Wilson et al. 1981). In this last study, a higher relative risk of cancer was seen in children whose mothers took drugs during pregnancy and were exposed to obstetric X-rays, than in children whose mothers took drugs but were not exposed to X-rays. A general difficulty with this type of study is the problem of obtaining accurate information about drug exposure during pregnancy, even when medical records relating to the pregnancy are available, and possible problems related to recall bias should be addressed. A study specifically designed to examine recall bias in mothers of children with life-threatening diseases is in progress, and preliminary analyses suggest that recall bias does not account for observed case-control differences (P. A. McKinney, pers. commn).

Postnatal exposures for benign conditions

Radiation

While the causative role of prenatal exposure to X-rays in the induction of leukaemia and solid tumours in children remains a matter of some controversy, the carcinogenic effect of postnatal therapeutic irradiation is well established. A number of studies have demonstrated an excess risk of leukaemia in children receiving moderate to heavy doses of therapeutic radiation (Stewart et al. 1958; Polhemus and Koch 1959; Murray et al. 1959; Graham et al. 1966). Furthermore, Bross and Natarajan (1974) reported that children with histories of allergic or infectious conditions were particularly susceptible to leukaemogenesis following therapeutic radiation, although this latter effect has not been confirmed in other studies.

Hempelmann et al. (1975) evaluated the risk of thyroid malignancy among persons who had been irradiated for thymic enlargement during infancy, and estimated the risk to be almost 100 times greater than in the general population. A dose-response relationship was apparent, with the greatest risk among children who received at least 4 Gy; the linear risk coefficient was estimated at 2.5 thyroid cancers per million person-years per rad to the thyroid.

A second historical cohort study showing increased risk of cancer following irradiation has involved follow-up of children receiving scalp irradiation to treat ringworm (tinea capitis). Nearly 11 000 children irradiated with a mean dose of 1.5 Gy to neural tissue were followed up, together with more than 16 000 unirradiated controls, from 1950 to 1972. The greatest cancer excesses were found for malignant neoplasms of the brain, parotid, and thyroid. The minimum interval between irradiation and diagnosis of malignancy was 4 years and the maximum 21 years, with a median interval of 11 years (Modan *et al.* 1974). The follow-up on this cohort of patients has been extended by Ron *et al.* (1988), confirming the earlier results. Sixty neural tumours were found in the exposed group, giving a relative risk compared with the controls of 6.9 for all tumours, and 8.4 for neural tumours of the head and neck. A marked dose-response relationship was observed and the relative risk approached 20-fold with doses greater than about 2.5 Gy. Higher risks were seen among males, persons of North African ethnic origin, and patients irradiated during early childhood. A smaller study of children irradiated for ringworm in New York City has produced broadly similar results (Shore *et al.* 1976).

A study of nearly 5000 patients irradiated in infancy for haemangioma of the skin showed no excess of cancer deaths (Li *et al.* 1974). The authors conclude that the apparent lack of carcinogenic effect may be due to small field sizes and relatively low radiation doses to internal organs.

Drugs

There are no reported associations between postnatal exposure to drugs and solid tumours in childhood. A population-based case-control study of childhood leukaemia, however, carried out in Shanghai, China, has demonstrated a significant dose-response relation between chloramphenicol usage and the risk of both acute lymphocytic leukaemia and acute non-lymphocytic leukaemia. Comparing usage in cases and controls, the odds ratio increased from 1.8 to 9.4, with successive categories of total days of chloramphenicol use from 1 to 5 days to more than 10 days. Furthermore, a significant risk of acute non-lymphocytic leukaemia was also found with use of syntomycin, which is pharmacologically related to chloramphenicol. Systemic chloramphenicol is now used only infrequently in paediatric practice in developed countries, but it is widely used in developing countries, and in the Shanghai study 34 per cent of cases and 17.6 per cent of controls reported having used chloramphenicol (Shu *et al.* 1988). Other studies are required to substantiate this important finding. The association between exposure to chloramphenicol in early childhood and an increased risk of leukaemia is particularly interesting in view of previous reports associating chloramphenicol with leukaemia (Mukherji 1957) and bone marrow depression (Fraumeni 1967).

Immunization

In the context of medical procedures which may influence carcinogenic risk, two recent reports which suggest that immunization against infectious diseases may have a *protective* effect against cancer are relevant. Kneale *et al.* (1986), reporting on an analysis of the OSCC data, found that cancer cases had had fewer immunizations against infectious diseases than their matched controls. The effect was seen for all types of immunizations and for all types of cancer. The case-control differences were greater among older cases with late immunizations than younger cases with early immunizations, and the effect was more pronounced for solid tumours than for leukaemias. The authors hypothesize that immunization, or 'simulated infection' affects the immune system such that further development of an *in situ* cancer is prevented or impeded.

The second study to report such an association, the Inter-Regional Epidemiological Study of Childhood Cancer (IRESCC) was also carried out in the UK (Hartly *et al.* 1988). Whereas the OSCC was based on a mortality series, the IRESCC study used incident cases. The methods of control selection in the two studies differed. Nevertheless, the results with respect to immunizations were very similar, and children who had never been immunized were found to be at significantly higher risk of developing any childhood cancer, compared with children who had received one or more immunizations (relative risk 3.58). The effect was found in all diagnostic groups and was not specific to any particular vaccine. Children less than two years at diagnosis were excluded from the analysis because symptoms due to onset of the tumour may have been the reason for the non-immunization in this group. These observations reported by two independent studies have important implications and merit further investigation.

Treatment for childhood malignancies

Survival from childhood cancer has improved dramatically over the past 30 years (Miller and McKay 1984; Birch *et al.* 1988). Long-term problems associated with cure from cancer are of greater potential importance in children who have the whole of their reproductive and economically productive lives in front of them, than in adults who develop cancer in later life. Children are growing rapidly and have a greater cell turnover than adults at the time of treatment, and may therefore be expected to be more sensitive to any carcinogenic effects of that treatment. Furthermore, as discussed above, genetic factors may play an important role in the histogenesis of many childhood cancers, and genetic predisposition may compound the inherent susceptibility of growing tissues to the potential carcinogenic effects of certain cancer treatments.

Overall risk of second malignancy

There have been two major multi-institution studies of second malignant neoplasms among survivors of childhood cancer. The prospective study coordinated by the Childhood Cancer Research Group in Oxford (CCRG) so far includes 90 second primary cancers diagnosed among about 10 000 3-year survivors from childhood cancer (Hawkins et al. 1987). The Late Effects Study Group (LESG) has compiled an international register of nearly 400 second malignant neoplasms in childhood cancer survivors, ascertained from 13 specialized North American and European institutions, each of which has a well-documented patient data base to provide a suitable denominator for calculation of risk (Meadows 1988).

Among children who had survived at least 2 years following diagnosis of their initial cancer in the LESG series, the cumulative risk up to age 25 years of developing a subsequent malignant neoplasm was 12.1 per cent (Tucker et al. 1984). The risk was found to vary according to the initial diagnosis, with lower risks following acute lymphocytic leukaemia, and higher risks following retinoblastoma, Wilms' tumour, soft tissue sarcomas, and in patients with cancer-prone syndromes such as neurofibromatosis. As discussed below, many of these second tumours could be directly attributed to previous radiotherapy or treatment with cytotoxic drugs, but a number of second malignancies were seen in children whose only previous treatment had been surgery. In these children genetic predisposition may have been the main aetiological factor. In other children, genetic and therapy-related factors may interact to produce a second malignancy. The most common first primary neoplasm in children subsequently developing other malignancies was retinoblastoma, followed in rank order by Hodgkin's disease, soft tissue sarcoma and Wilms' tumour. The most common second malignant neoplasms were sarcomas of bone, followed in rank order by soft tissue sarcoma, leukaemias and lymphomas, and brain tumours. The most common first–second tumour associations were retinoblastoma followed by osteosarcoma or soft tissue sarcoma, and Hodgkin's disease followed by leukaemia (Meadows et al. 1985).

Among children surviving at least three years after diagnosis of a first primary cancer in the CCRG study, the cumulative risk of developing a second malignancy before the age of 25 was about 4 per cent, six times greater than the cancer risk in the general population. The most common first primary tumours in this series were of the central nervous system, followed in rank order by retinoblastoma, acute leukaemia, and Wilms' tumour. The most common second primaries were osteosarcoma, followed in rank order by brain tumours, leukaemia, and soft tissue sarcoma. In common with the LESG, the most frequent combination of first and second malignancies was retinoblastoma followed by osteosarcoma (Hawkins et al. 1987; Kingston et al. 1987).

In some cases, the particular combinations of first and second tumours seen in these studies may be due to a carcinogenic effect of the treatment for the first tumour which is specific for a particular site or type of subsequent primary. In other cases, specific associations of cancers may reflect genetic predisposition to both tumours, or else the enhanced carcinogenicity of cytotoxic treatment in individuals who are genetically susceptible.

The CCRG study in Britain comprises a population-based series, whereas most institutions in the LESG international study contributed a hospital-based case series. The differences in risk estimates and in the observed patterns of multiple primary cancer seen in the two studies, discussed below, may be due to such differences in the patient population and in the treatments given for first primary tumours, particularly radiation doses.

Radiation and chemotherapy

In the CCRG study, 60 per cent of the second malignancies were thought to be radiation-associated, i.e. solid tumours developing within or on the edge of radiation treatment fields, or leukaemias developing in children who had previously received radiotherapy. The comparable figure from the LESG is 67 per cent. Chemotherapy alone (without radiotherapy) as treatment for the first primary was associated with 10 per cent of the subsequent malignancies in the CCRG study, and with 17 per cent in the LESG study. In these two studies, however, 30 and 16 per cent, respectively, of the second malignancies occurred in children who had received no chemotherapy and no radiotherapy, or who had developed a second tumour at a site distant from the radiation treatment field. These second tumours must be assumed to represent an inherent, biological increase in tumour risk associated with some first primary tumours, such as osteosarcoma following retinoblastoma.

Table 6.2 shows the pattern of second malignancies judged to be associated or not associated with the treatment given for the first primary malignancy in the two series. In both studies bone sarcoma, mainly osteosarcoma, was principally associated with prior radiotherapy.

The LESG has conducted both case-control and cohort analyses to examine the relationship between therapy for a malignancy in childhood and subsequent development of bone sarcoma. Sixty-four patients with a second primary bone cancer were compared to 209 matched controls who had survived an initial cancer without development of a subsequent neoplasm. The risk of bone sarcoma depended on the total radiation dose to the site at which the bone sarcoma subsequently developed. There was a sharp dose-response gradient, rising to about 40-fold after doses to the bone of greater than 60 Gy. The patterns of risk among dose categories did not differ between ortho-voltage and mega-voltage radiation. The risk of bone sarcoma was also increased 5-fold by alkylating agent chemotherapy, independent of radiation therapy, and this risk increased with cumulative drug exposure.

In this study, cases and controls with bilateral retinoblastoma were matched, in order to remove the influence of genetic factors associated

Table 6.2 Association of second malignancy with treatment for first malignancy

Type of second malignancy	Number of second malignancies		Per cent associated with radiotherapy		Per cent associated with chemotherapy		Per cent not associated with either radiotherapy or chemotherapy	
	CCRG[1]	LESG[2]	CCRG	LESG	CCRG	LESG	CCRG	LESG
Bone sarcoma	35	57	49	78	17	12	34	10
Soft tissue sarcoma	17	59	47	73	6	10	47	17
Leukaemia/lymphoma	23	59	74	61	22	20	4	19
Skin cancers	18	31	78	61	0	16	22	23
Central nervous system tumours	29	29	76	45	3	38	21	17
Thyroid cancers	4	26	75	92	0	4	25	4
Other tumours	25	37	40	57	8	16	52	27
Total	151	308	60	67	10	17	30	16

[1] Childhood Cancer Research Group (Kingston et al. 1987).
[2] Late Effects Study Group (Meadows et al. 1985).

with retinoblastoma. The relative risk of new bone sarcomas in each radiotherapy dose category was similar for retinoblastoma and all other tumours. In other words, most of the absolute excess risk of bone sarcoma in retinoblastoma patients is due to genetic factors. Similarly, after controlling for genetic factors, an excess of bone sarcoma associated with alkylating agent therapy was observed (Tucker et al. 1987a). Analysis of the CCRG retinoblastoma series suggests that cyclophosphamide may potentiate the effect of radiation therapy in the induction of subsequent osteosarcomas (Draper et al. 1986).

A cohort analysis of over 9000 patients who had survived cancer in childhood for 2 or more years was also carried out, to estimate the absolute risk of developing a second primary bone cancer. In this study, 48 cases of bone cancer were observed, compared to only 0.4 expected (relative risk 133). The absolute excess risk was estimated to be 9.4 cases per 10 000 persons per year. The risk was highest among children previously treated for retinoblastoma, whose risk of osteosarcoma was increased 1000-fold (relative risk 999, excess risk 53.6 per 10 000 per year) and among children treated for Ewing's sarcoma (relative risk 649, excess risk 59.6 per 10 000 per year). Relative risk increased significantly with time since treatment, and the overall cumulative risk of bone cancer 20 years after an initial cancer diagnosis was 2.8 per cent for all first cancers, and 14.1 per cent for retinoblastoma.

Leukaemia was another common second malignancy, associated with both radiotherapy and chemotherapy. Leukaemia risk was evaluated in the LESG cohort of more than 9000 children referred to above, and a case-control study was carried out in 25 cases of leukaemia occurring in childhood cancer survivors, and 90 matched control children who had survived childhood cancer without developing leukaemia (Tucker et al. 1987b). Twenty-two leukaemias occurred among the cohort, compared with 1.52 expected (relative risk 14). The risk of leukaemia rose with increasing age at diagnosis of the first cancer. The highest leukaemia risks were seen in patients with Hodgkin's disease, Wilms' tumour, and Ewing's tumour as their first malignancy. All patients who developed leukaemia had received either radiotherapy or treatment with alkylating agents. Total dose of all alkylating agents was estimated for each patient and expressed as an alkylating agent 'score'. Leukaemia risk increased with increasing alkylating agent score, and in the highest dose category the relative risk of leukaemia was 23. With respect to leukaemia risk, the total dose of alkylating agent was found to be more important than the duration of treatment. Doxorubicin was the only other chemotherapeutic agent which appeared to be associated with leukaemia risk, and patients who had received alkylating agents and doxorubicin were at much greater risk of leukaemia (relative risk more than 40). After adjustment for alkylating agent treatment, no difference was found in leukaemia risk by average radiation dose to the bone marrow, and in those patients receiving radiotherapy alone, there was no evidence of any dose response. The authors conclude that the excess risk of leukaemia following treatment for cancer in childhood was almost entirely due to alkylating agent therapy.

Thyroid cancers as second malignancies occurred overwhelmingly in children who had received prior radiotherapy, and 27 out of the 30 thyroid cancers occurring in the CCRG and LESG series combined were radiation-associated. The LESG found that the risk of thyroid cancer increased with time since first cancer, and even after 20 years there was no plateau. The highest risks were seen in children whose first cancer was diagnosed between birth and 4 years of age, and—in contrast to the age-dependence of leukaemia risk (see above)—the risk of a subsequent thyroid cancer decreased with increasing age at diagnosis of the first cancer. Neuroblastoma, Hodgkin's disease and Wilms' tumour as first cancers were associated with the greatest risk of second primary thyroid cancer. There was a marked dose-response relationship, with a maximum 12-fold risk in patients who had received more than 60 Gy to the thyroid (Tucker *et al.* 1986). In this study, the risk of thyroid cancer in patients receiving over 60 Gy was compared with that in patients receiving less than 2 Gy. Other workers have demonstrated that this level of radiation was associated with an approximately 10-fold increase in the risk of thyroid cancer (Shore *et al.* 1984). Thus exposure to more than 60 Gy may increase the risk of thyroid cancer by around 100-fold, compared with population rates. A recent report from Villejuif, France, on thyroid cancer following radiotherapy for childhood malignancy confirms the work of Tucker *et al.* (1986) and further highlights the apparent sensitivity of neuroblastoma patients to subsequent development of thyroid cancer. The risk of thyroid cancer in children treated for neuroblastoma, adjusted for age at diagnosis of first cancer, sex, chemotherapy, and radiotherapy dose to the thyroid, was 7.2-fold higher than in children treated for other types of first primary tumour (de Vathaire *et al.* 1988).

Kushner *et al.* (1988) conducted a study of second malignancies among 254 previously untreated patients diagnosed with Hodgkin's disease during childhood. Twelve second malignancies occurred, compared with 0.61 expected (RR = 19.8, 95 per cent CI 10.2–34.6). Among patients who received multiple agent chemotherapy the cumulative risk of second malignancy at 15 years was 19 per cent, whereas in patients who received either single-agent chemotherapy or radiation alone the cumulative risk was only 2 per cent. The risk of acute non-lymphocytic leukaemia in this patient group was highest during the period 5 to 10 years after combined modality treatment. Bone sarcomas were most commonly seen during the first 10 years after treatment, and carcinomas, for example breast and colon, were seen later in life. The authors conclude that there is an enhanced risk of solid second malignancies related to therapy, but that these tumours develop at approximately the normal age.

In a follow-up study of nearly 2500 patients included in the US National Wilms' Tumour Study, Breslow *et al.* (1988) found 15 second malignant neoplasms compared with 1.8 expected (relative risk 8.5). The cumulative risk of developing a second malignancy among these patients was 1 per cent up to 10 years after diagnosis of Wilms' tumour, and higher rates were seen in those patients receiving radiation as part of their initial treatment, but the difference did not reach statistical significance.

Cancer in the offspring of survivors

In experimental systems both radiation and alkylating agents can be mutagenic. The possibility arises that treatment of childhood malignancy with such agents may result in germ cell mutations, which might be expressed as childhood malignancies in the offspring of individuals treated for cancer in childhood. Two recent studies (Mulvihill *et al.* 1987; Li *et al.* 1987) of the offspring of childhood cancer survivors did not demonstrate such an effect, but so far the number of childhood cancer survivors treated with multiple-modality therapy (and for whom details of the offspring are available) is small, and long-term surveillance of this potential hazard will be necessary.

Genetic factors

It is probable that in a substantial proportion of childhood cancers, genetic factors are of aetiological importance (Knudson 1976). Children who have survived one tumour may be genetically susceptible to the development of further malignancies, and this susceptibility may render them particularly sensitive to the carcinogenic effects of treatments with radiation and certain drugs, for example alkylating agents. Some patients may have clearly defined genetic disease, for example Gorlin's syndrome and neurofibromatosis, but in others genetic susceptibility may manifest itself as familial clusters of cancer, as in the Li–Fraumeni cancer family syndrome (Li *et al.* 1988).

Genetic or bilateral retinoblastoma has been the childhood cancer most extensively studied with respect to the relationship between treatment and development of second primary cancers. Retinoblastoma is a particularly good model in which to carry out this type of investigation for the following three reasons: first, it has had an excellent prognosis for a considerable number of years; second, the observed pattern of inheritance is consistent with a single dominant gene with high penetrance; and third, gene carriers are easily identified. Second primary tumours following treatment for bilateral retinoblastoma were reported as early as 1949 (Reese *et al.* 1949), and a number of hospital-based series, for example Abramson *et al.* (1976) and Kitchin and Elsworth (1974), have suggested that retinoblastoma patients are at very high risk of developing subsequent primaries, particularly osteosarcoma.

Hospital-based cancer series may well overestimate the risk of second malignancy, however, because healthy patients are more likely to be lost to follow-up than patients who have problems. The availability of two recent population-based surveys among retinoblastoma patients thus enables a more accurate estimate of second cancer risk to be obtained. Draper *et al.* (1986) reported on a population-based series of 882 retinoblastoma patients in Britain, among whom 384 were known to have the genetic form. Of 30 patients who developed second primary tumours, 26 had had genetic retinoblastoma. In these 26 patients, 12 of the second primaries developed outside radiation treatment fields. Considering only osteosarcomas occurring outside the treatment field, an inherent risk of developing second primary osteosarcoma was estimated to be 2.2 per cent by 18 years, which is approximately 200 times that for the general population. Eight osteosarcomas occurred within the radiation field, in the orbit, maxilla, and nasal bones. Cancers at these sites are very rare, and less than one person in 100 000 would be expected to develop osteosarcoma at any site in the skull and jawbones under the age of 20. Overall, the estimated cumulative incidence rate for second primary tumours within the radiation field in patients with genetic retinoblastoma was 6.6 per cent by 18 years, and 3.7 per cent for osteosarcomas alone. Draper *et al.* (1986) concluded that there was evidence to suggest that genetic retinoblastoma patients are particularly sensitive to radiogenic tumours.

The second population-based study utilized all patients included in a national retinoblastoma registry in the Netherlands during 1945–1970 (Der Kinderen *et al.* 1988). Among 393 patients, 141 had hereditary retinoblastoma. All 17 patients who developed a second primary cancer had had genetic retinoblastoma. The cumulative incidence of non-ocular cancer in the genetic retinoblastoma cases was 19 per cent at 35 years, compared with 1.3 per cent in the general population of the Netherlands. Eight patients developed non-ocular cancers within the radiation treatment fields used for retinoblastoma, whereas among the patients who did not receive radiotherapy, none developed a cancer in the same area. The cumulative incidence of radiation-induced non-ocular cancer was estimated to be 11 per cent at 35 years.

It may be concluded from these two studies that survivors of genetic retinoblastoma are at a particularly high risk of developing second primaries, notably osteosarcoma, but that the risk is further increased by exposure to radiation. Survivors of non-hereditary retinoblastoma do not appear to be at particularly high risk.

Perhaps the most dramatic example of the interaction between treatment of a first primary cancer with radiotherapy and development of subsequent primary cancers is seen in patients with Gorlin's syndrome

(Gorlin and Goltz 1960). This syndrome is characterized by multiple basal cell carcinomas, jaw cysts, skeletal deformities, skin pits on palms and soles, soft tissue calcification, occasionally mental retardation, and other neoplasms including medulloblastoma and ovarian fibromas. Patients can develop more than 100 basal cell carcinomas spontaneously during their lifetime, but radiation may result in the rapid onset of such lesions. This effect is seen characteristically in children treated with craniospinal irradiation for medulloblastoma, and in whom the syndrome may not have been recognized at the time of diagnosis (Fig. 6.2). Large numbers of basal cell carcinomas develop within the radiation treatment field, but especially at

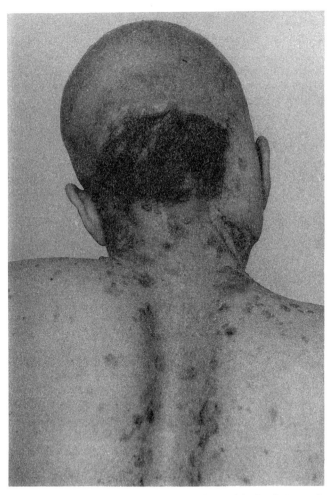

Fig. 6.2 Multiple basal cell carcinomas in a patient with Gorlin's syndrome who had previously received craniospinal irradiation for medulloblastoma.

the edge of it, and these can appear within months of radiotherapy (Strong 1977).

Retinoblastoma and Gorlin's syndrome are well-defined heritable conditions. Other conditions predisposing to multiple primary cancers include von Recklinghausen's neurofibromatosis and multiple endocrine neoplasia (Sippel's syndrome). Less well-defined are the cancer family syndromes. Among the CCRG series described by Kingston et al. (1987), 53 (35 per cent) of the 151 patients with multiple primaries in the series had recognized genetic disease, and a family history of malignancy in at least one first degree relative was recognized in eight other cases.

In the LESG series described by Meadows et al. (1985), 23 patients with double primary cancers had siblings who had also had a similar or identical cancer in childhood. Nine sibling pairs each had retinoblastoma, six pairs each had leukaemia or lymphoma, four had brain tumours, two had Wilms' tumours, and two had soft tissue sarcomas. In five families, there were three or more siblings with childhood cancer, and six patients had siblings with more than one primary malignancy.

Farwell and Flannery (1984) reported on a series of children who developed second primary malignancies following central nervous system tumours. In this series nine out of 670 children with central nervous system tumours ascertained from the Connecticut tumour registry developed second primary tumours, giving a relative risk of 9.1 compared with the number of tumours expected from age-specific incidence rates in the general population. Five cancers were found among the first degree relatives of these nine patients compared with 0.9 expected, and the authors concluded that children with central nervous system tumours are not only at increased risk for other cancers, but that children with CNS tumours and second cancers may be part of a familial cluster with increased risk of cancer.

Strong et al. (1987) carried out a survey of 159 3-year survivors of childhood soft tissue sarcoma and their relatives, in order to determine the incidence of second malignancies in the patients and of first cancer in their relatives. Eight of the index patients developed a second malignancy, compared with less than one expected from population-based rates. There was also a significant excess of cancers among the first-degree relatives of the index patients, notably of soft tissue, bone, and breast. These observations are consistent with the familial cancer syndrome described by Li and Fraumeni (1969) and Li et al. (1988). Familial cancer risk was determined in a number of sub-groups of patients defined according to various characteristics of the index patient. A highly significant excess of cancer was found among the relatives of patients with second malignancies. Furthermore, the types of tumour which occurred in excess in close relatives were also found as second malignancies in the index patients.

These findings have important clinical implications. Clustering of cancers in the family of a child with cancer may indicate that that child has an elevated risk of developing a subsequent primary, and may be particularly sensitive to the carcinogenic effects of radiotherapy and chemotherapy. Conversely, the development of a second primary malignancy in a child may be the result of genetic predisposition, and the close relatives of that child may themselves be at high risk of developing malignant disease, even in the absence of familial clustering of cancer at the time of the child's diagnosis.

Conclusion

The overall contribution of medical exposures to the induction and development of childhood cancers in general is small, especially in comparison to the number of children whose lives can be saved by therapy, but the contribution to our understanding of the causes of childhood cancer which may be gained from the study of cancer risks following medical treatment cannot be understated. Whether or not prenatal exposure to diagnostic X-rays is causally associated with subsequent childhood cancer is now largely academic, since ultrasound has largely replaced obstetric X-rays as a diagnostic tool, but the resolution of this problem would still be important in the current debate on the role of low-level environmental radiation in carcinogenesis.

Long-term survival from childhood cancer is no longer unusual, and study of the late effects of treatment, particularly the occurrence of second primary malignancies, has profound implications for the design of future treatment protocols in both children and adults (see Fraser, Chapter 8, this volume). Children who develop second primary malignancies may do so as a consequence of genetic susceptibility, but the occurrence of multiple primary cancers should alert the clinician to the possibility of underlying genetic disease in the family, and close relatives of the child may also be at high risk of cancer.

In addition to the clinical implications of delayed cancer risk following medical treatment, there is also a great deal still to be learned from studies of the carcinogenicity of medical treatments in children, both about the carcinogenic process in general and especially about interactions between genetic and environmental factors in the development of cancer.

Acknowledgements

The Manchester Children's Tumour Registry is supported by the Cancer Research Campaign.

References

Abramson, D. H., Ellsworth, R. M., and Zimmerman, L. E. (1976). Non-ocular cancer in retinoblastoma survivors. *Ophthalmology*, **81**, 454–7.

Beebe, G. W., Kato, H., and Land, C. E. (1978). Studies of the mortality of A-bomb survivors. 6. Mortality and radiation dose, 1950–1974. *Radiation Research*, **75**, 138–201.

Bibbo, M., Haenszel, W. M., Wied, G. L., Hubby, M., and Herbst, A. L. (1978). A twenty-five year follow-up study of women exposed to diethylstilbestrol during pregnancy. *New England Journal of Medicine*, **298**, 763–7.

Birch, J. M., Marsden, H. B., Morris Jones, P. H., Pearson, D., and Blair, V. (1988). Improvements in survival from childhood cancer: results of a population-based survey over 30 years. *British Medical Journal*, **296**, 1372–6.

Breslow, N. E., Norkool, P. A., Olshan, A., Evans, A., and D'Angio, G. (1988). Second malignant neoplasms in survivors of Wilms' tumor: a report from the National Wilms' Tumor Study. *Journal of the National Cancer Institute*, **80**, 592–5.

Brian, D. D., Tilley, B. C., Labarthe, D. R., O'Fallon, J. W., Noller, K. L., and Kurland, L. T. (1980). Breast cancer in DES-exposed mothers: absence of association. *Mayo Clinic Proceedings*, **55**, 89–93.

Bross, I. D. J. and Natarajan, N. (1974). Risk of leukemia in susceptible children exposed to preconception, *in utero*, and postnatal radiation. *Preventive Medicine*, **3**, 361–9.

Burch, P. R. J. (1970). Prenatal radiation exposure and childhood cancer. *Lancet*, **ii**, 1189.

Burch, P. R. J. (1974). Radiology now—the 10-day rule. *British Journal of Radiology*, **47**, 198.

Cavenee, W. K., et al. (1983). Expression of recessive alleles by chromosomal mechanisms in retinoblastoma. *Nature*, **305**, 779–84.

Court Brown, W. M., Doll, R., and Hill, A. B. (1960). Incidence of leukaemia after exposure to diagnostic radiation *in utero*. *British Medical Journal*, **ii**, 1539–45.

Der Kinderen, D. J., Koten, J. W., Nagelkerke, N. J. D., Tan, K. E. W. P., Beemer, F. A., and Den Otter, W. (1988). Non-ocular cancer in patients with hereditary retinoblastoma and their relatives. *International Journal of Cancer*, **41**, 499–504.

de Vathaire, F., Francois, P., Schweisguth, O., Oberlin, O., and Le, M. G. (1988). Irradiated neuroblastoma in childhood as potential risk factor for subsequent thyroid tumour. *Lancet*, **ii**, 455.

Draper, G. J., Sanders, B. M., and Kingston, J. E. (1986). Second primary neoplasms in patients with retinoblastoma. *British Journal of Cancer*, **53**, 661–71.

Druckrey, H., Ivankovic, S., and Preussman, R. (1966). Teratogenic and carcinogenic effects in the offspring after single injection of ethylnitrosourea to pregnant rats. *Nature*, **210**, 1378–9.

Farwell, J. and Flannery, J. T. (1984). Cancer in relatives of children with central-nervous-system neoplasms. *New England Journal of Medicine*, **311**, 749–53.

Fraumeni, J. F. Jr (1967). Bone marrow depression induced by chloramphenicol or phenylbutazone. *Journal of the American Medical Association*, **201**, 828–34.

Gold, E., Gordis, L., Tonascia, J., and Szklo, M. (1978). Increased risk of brain

tumors in children exposed to barbiturates. *Journal of the National Cancer Institute*, **61**, 1031–4.
Gorlin, R. J. and Goltz, R. W. (1960). Multiple nevoid basal cell epithelioma, jaw cysts and bifid ribs: a syndrome. *New England Journal of Medicine*, **262**, 908–12.
Graham, S., et al. (1966). Preconception, intrauterine, and postnatal irradiation as related to leukemia. *National Cancer Institute Monograph*, **19**, 347–71.
Hanson, J. W. and Smith, D. W. (1975). The fetal hydantoin syndrome. *Journal of Pediatrics*, **87**, 285–90.
Hartley, A. L., et al. (1988). The inter-regional epidemiological study of childhood cancer (IRESCC): past medical history in children with cancer. *Journal of Epidemiology and Community Health*, **42**, 235–42.
Harvey, E. B., Boice, J. D. Jr, Honeyman, M., and Flannery, J. T. (1985). Prenatal X-ray exposure and childhood cancer in twins. *New England Journal of Medicine*, **312**, 541–5.
Hawkins, M. M., Draper, G. J., and Kingston, J. E. (1987). Incidence of second primary tumours among childhood cancer survivors. *British Journal of Cancer*, **56**, 339–47.
Heinonen, O. P. (1973). Diethylstilbestrol in pregnancy: frequency of exposure and usage patterns. *Cancer*, **31**, 573–7.
Hempelmann, L. H., Hall, W. J., Phillips, M., Cooper, R. A., and Ames, W. R. (1975). Neoplasms in persons treated with X-rays in infancy: fourth survey in 20 years. *Journal of the National Cancer Institute*, **55**, 519–30.
Herbst, A. L. (1981). Clear cell adenocarcinoma and the current status of DES-exposed females. *Cancer*, **48**, 484–8.
Herbst, A. L. and Scully, R. E. (1970). Adenocarcinoma of the vagina in adolescence. *Cancer*, **25**, 745–57.
Herbst, A. L., Ulfelder, H., and Poskanzer, D. C. (1971). Adenocarcinoma of the vagina: association of maternal stilbestrol therapy with tumor appearance in young women. *New England Journal of Medicine*, **284**, 878–81.
Herbst, A. L., Cole, P., Colton, T., Robboy, S. J., and Scully, R. E. (1977). Age-incidence and risk of diethylstilbestrol-related clear cell adenocarcinoma of the vagina and cervix. *American Journal of Obstetrics and Gynaecology*, **128**, 43–50.
Herbst, A. L., Scully, R. E., and Robboy, S. J. (1979). Prenatal diethylstilbestrol exposure and human genital tract abnormalities. In *Perinatal carcinogenesis*, (ed. F. I. Gregoric), NCI Monograph 51, DHEW Publication No. (NIH) 79-1633, pp. 25–35. National Cancer Institute, Bethesda.
Innis, M. D. (1972). Nephroblastoma: possible index cancer of childhood. *Medical Journal of Australia*, **1**, 18–20.
Jablon, S. and Kato, H. (1970). Childhood cancer in relation to prenatal exposure to atomic-bomb radiation. *Lancet*, **ii**, 1000–3.
Kingston, J. E., Hawkins, M. M., Draper, G. J., Marsden, H. B., and Kinnier Wilson, L. M. (1987). Patterns of multiple primary tumours in patients treated for cancer during childhood. *British Journal of Cancer*, **56**, 331–8.
Kinnier Wilson, L. M., Kneale, G. W., and Stewart, A. M. (1981). Childhood cancer and pregnancy drugs. *Lancet*, **ii**, 314–15.
Kitchin, F. D., and Ellsworth, R. M. (1974). Pleiotropic effects of the gene for retinoblastoma. *Journal of Medical Genetics*, **11**, 244–6.
Kneale, G. W., Stewart, A. M., and Kinnier Wilson, L. M. (1986). Immunizations

against infectious diseases and childhood cancers. *Cancer Immunology and Immunotherapy*, **21**, 129–32.

Knudson, A. G. Jr (1976). Genetics and the etiology of childhood cancer. *Pediatric Research*, **10**, 513–17.

Kramer, S., Ward, E., Meadows, A. T., and Malone, K. E. (1987). Medical and drug risk factors associated with neuroblastoma: a case-control study. *Journal of the National Cancer Institute*, **78**, 797–804.

Kushner, B. H., Zauber, A., and Tan, C. T. C. (1988). Second malignancies after childhood Hodgkin's disease. *Cancer*, **62**, 1364–70.

Li, F. P. and Fraumeni, J. F. Jr (1969). Soft-tissue sarcoma, breast cancer and other neoplasms. A familial syndrome? *Annals of Internal Medicine*, **71**, 747–52.

Li, F. P., Cassady, J. R., and Barnett, E. (1974). Cancer mortality following irradiation in infancy for hemangioma. *Radiology*, **113**, 177–8.

Li, F. P., *et al.* (1987). Outcome of pregnancy in survivors of Wilms' tumor. *Journal of the American Medical Association*, **257**, 216–19.

Li, F. P., *et al.* (1988). A cancer family syndrome in twenty-four kindreds. *Cancer Research*, **48**, 5358–62.

Lipson, A. and Bale, P. (1985). Ependymoblastoma associated with prenatal exposure to diphenylhydantoin and methylphenobarbitone. *Cancer*, **55**, 1859–62.

MacMahon, B. (1962). Prenatal X-ray exposure and childhood cancer. *Journal of the National Cancer Institute*, **28**, 1173–91.

MacMahon, B. (1985). Prenatal X-ray exposure and twins. *New England Journal of Medicine*, **312**, 576–7.

MacMahon, B. and Hutchison, G. B. (1964). Prenatal X-ray and childhood cancer: a review. *Acta UICC*, **20**, 1172–4.

Majewski, F., Steger, M., Richter, B., Gill, J., and Rabe, F. (1981). The teratogenicity of hydantoins and barbiturates in humans. *International Journal of Biological Research in Pregnancy*, **2**, 37–45.

McKinney, P. A., *et al.* (1987). The inter-regional epidemiological study of childhood cancer (IRESCC): a case-control study of aetiological factors in leukaemia and lymphoma. *Archives of Disease in Childhood*, **62**, 279–87.

Meadows, A. T. (1988). Risk factors for second malignant neoplasms: report from the Late Effects Study Group. *Bulletin du Cancer*, **75**, 125–30.

Meadows, A. T., *et al.* (1985). Second malignant neoplasms in children: an update from the Late Effects Study Group. *Journal of Clinical Oncology*, **3**, 532–8.

Miller, R. W. (1969). Delayed radiation effects in atomic-bomb survivors. *Science*, **166**, 569–74.

Miller, R. W. and McKay, F. M. (1984). Decline in U.S. childhood mortality 1959 through 1980. *Journal of the American Medical Association*, **251**, 1567–70.

Modan, B., Baidatz, D., Mart, H., Steinitz, R., Levin, S. G. (1974). Radiation-induced head and neck tumours. *Lancet*, **i**, 277–9.

Mole, R. H. (1974). Antenatal irradiation and childhood cancer: causation or coincidence? *British Journal of Cancer*, **30**, 199–208.

Monson, R. R. and MacMahon, B. (1984). Prenatal X-ray exposure and cancer in children. In *Radiation carcinogenesis: epidemiology and biological significance*, (ed. J. D. Boice, Jr and J. F. Fraumeni, Jr), pp. 97–105. Raven Press, New York.

Mukherji, P. S. (1957). Acute myeloblastic leukaemia following chloramphenicol treatment. *British Medical Journal*, **i**, 1286–7.

Mulvihill, J. J., et al. (1987). Cancer in offspring of long-term survivors of childhood and adolescent cancer. *Lancet*, **ii**, 813–17.

Murray, R., Heckel, P., and Hempelmann, L. H. (1959). Leukemia in children exposed to ionizing radiation. *New England Journal of Medicine*, **261**, 585–9.

Parkin, D. M., Stiller, C. A., Draper, G. J., Bieber, C. A., Terracini, B., and Young, J. L. (ed.) (1988a). *International incidence of childhood cancer*, IARC Scientific Publications No. 87. International Agency for Research on Cancer, Lyon.

Parkin, D. M., Stiller, C. A., Draper, G. J., and Bieber, C. A. (1988b). The international incidence of childhood cancer. *International Journal of Cancer*, **42**, 511–20.

Polhemus, D. W. and Koch, R. (1959). Leukemia and medical radiation. *Pediatrics*, **23**, 453–61.

Preston-Martin, S., Yu, M. C., Benton, B., and Henderson, H. E. (1982). N-nitroso compounds and childhood brain tumors: a case-control study. *Cancer Research*, **42**, 5240–5.

Reese, A. G., Merriam, G. R., and Martin, H. E. (1949). Treatment of bilateral retinoblastoma by irradiation and surgery. Report on 15-year results. *American Journal of Ophthalmology*, **32**, 175–90.

Riccardi, V. M., Sujansky, E., Smith, A. C., and Francke, U. (1978). Chromosomal imbalance in the aniridia-Wilms' tumor association: 11p interstitial deletion. *Pediatrics*, **61**, 604–10.

Robison, L. L., et al. (1984). Down syndrome and acute leukemia in children: a 10-year retrospective survey from Children's Cancer Study Group. *Journal of Pediatrics*, **105**, 235–42.

Ron, E., et al. (1988). Tumors of the brain and nervous system after radiotherapy in childhood. *New England Journal of Medicine*, **319**, 1033–9.

Shore, R. E., Albert, R. E., and Pasternack, B. R. (1976). Follow up study of patients treated by X-ray epilation for tinea capitis: resurvey of post-treatment illness and mortality experience. *Archives of Environmental Health*, **31**, 21–8.

Shore, R. E., Woodward, E. D., and Hempelmann, L. H. (1984). Radiation-induced thyroid cancer. In *Radiation carcinogenesis: epidemiology and biological significance* (ed. J. D. Boice, Jr and J. F. Fraumeni, Jr), pp. 131–41. Raven Press, New York.

Shu, X. O., et al. (1988). A population-based case-control study of childhood leukemia in Shanghai. *Cancer*, **62**, 635–44.

Sotelo-Avila, C., Gonzalez-Crussi, F., and Fowler, J. W. (1980). Complete and incomplete forms of Beckwith–Wiedemann syndrome: their oncogenic potential. *Journal of Pediatrics*, **96**, 47–50.

Stewart, A. and Kneale, G. W. (1970). Radiation dose-effects in relation to obstetric X-rays and childhood cancers. *Lancet*, **i**, 1185–8.

Stewart, A. M., Webb, J. W., Giles, B. D., and Hewitt, D. (1956). Malignant disease in childhood and diagnostic irradiation *in utero*. *Lancet*, **ii**, 447.

Stewart, A., Webb, J., and Hewitt, D. (1958). A survey of childhood malignancies. *British Medical Journal*, **i**, 1495–508.

Strong, L. C. (1977). Theories of pathogenesis: mutation and cancer. In *Genetics of human cancer*, (ed. J. J. Mulvihill, R. W. Miller, and J. F. Fraumeni, Jr), pp. 401–15. Raven Press, New York.

Strong, L., Stine, E., and Norsted, T. L. (1987). Cancer in survivors of childhood soft tissue sarcoma and their relatives. *Journal of the National Cancer Institute*, **79**, 1213–20.

Totter, J. R. and MacPherson, H. G. (1981). Do childhood cancers result from prenatal X-rays? *Health Physics*, **40**, 511–24.

Tucker, M. A., Meadows, A. T., Boice, J. D. Jr, Hoover, R. N., and Fraumeni, J. F. Jr (1984). Cancer risk following treatment of childhood cancer. In *Radiation carcinogenesis: epidemiology and biological significance*, (ed. J. D. Boice, Jr and J. F. Fraumeni, Jr), pp. 211–24. Raven Press, New York.

Tucker, M. A., Meadows, A. T., Morris Jones, P., and Stovall, M. (1986). Therapeutic radiation at young age linked to secondary thyroid cancer. *Proceedings of the American Society of Clinical Oncology*, **5**, 211.

Tucker, M. A., *et al.* (1987a). Bone sarcomas linked to radiotherapy and chemotherapy in children. *New England Journal of Medicine*, **317**, 588–93.

Tucker, M. A., *et al.* (1987b). Leukemia after therapy with alkylating agents for childhood cancer. *Journal of the National Cancer Institute*, **78**, 459–64.

UNSCEAR (1977). *Sources and effects of ionizing radiation*, Publication E 771x1. United Nations Scientific Committee on the Effects of Atomic Radiation, New York.

Wechsler, W., Rice, J. M., and Vesselinovitch, S. D. (1979). Transplacental and neonatal induction of neurogenic tumours in mice: comparison with related species and with human pediatric neoplasms. In *Perinatal carcinogenesis*, (ed. F. I. Gregoric), NCI Monograph 51, DHEW Publication No. (NIH) 79–1633, pp. 219–26. National Cancer Institute, Bethesda.

Yoshimoto, Y., Kato, H., and Schull, W. J. (1988). Risk of cancer among children exposed *in utero* to A-bomb radiations 1950–1984. *Lancet*, **ii**, 665–9.

7
Surgery

MICHEL P. COLEMAN

Introduction

Surgery has been the mainstay of cancer treatment for centuries (Hill 1979). Elective surgery became possible with the advent of general anaesthesia and antisepsis in the 19th century, greatly extending the range of curative or radical surgery, both for cancer and non-malignant disease, while radiotherapy became available in the early 1900s, and chemotherapy has only developed since about 1950. For cancer, surgery also has major applications in diagnosis, palliative treatment, and the management of complications and intractable pain, as well as in reconstructive and plastic surgery (Raven 1984).

The short-term risk of illness and death caused by anaesthetic and surgical complications is well known, generally small and often predictable (Rosenberg 1989). Such risks are readily included in the surgeon's calculation of risk and benefit when deciding whether to operate. In comparison to the adverse effects of external agents such as radiotherapy and chemotherapy, the possibility that a surgical procedure might eventually cause cancer is harder to imagine, and difficult to incorporate into clinical decision-making (Vennin *et al.* 1985). But surgery often involves complete or partial removal of an organ, and may produce permanent changes in the anatomical arrangements and physiological functions which are not necessarily the primary objective of treatment. The evidence that some of these changes may increase—or reduce—subsequent cancer risk is reviewed in this chapter.

Primary prevention

Surgery is of value for the primary prevention of cancer in a few conditions, mostly rare, which are associated with a high and predictable risk. Colectomy is usually recommended in long-standing and extensive ulcerative colitis and in familial polyposis coli, in order to avoid the associated high risk of colon cancer. Orchidopexy or orchidectomy can be done to reduce or avoid the high risk of testicular cancer in a cryptorchid testis, and prophylactic thyroidectomy may be of value in selected patients with multiple endocrine neoplasia in the prevention of medullary carcinoma of the thyroid (Rosenberg 1989).

Partial gastrectomy

Cancer of the gastric remnant

The issue of whether partial gastrectomy for benign peptic ulceration is a cause of cancer in the stomach remnant has been the subject of a very large number of studies and reviews. Schmähl et al. (1977) collected over 300 case reports from the literature, and Penn (1982) mentions 1200 cases. Surgical procedures to treat both duodenal and gastric ulcer have been common for many years, but there are relatively few studies in which the number of gastric cancers observed can be compared to an expected number derived from a suitable comparison population. The results of six large cohort studies are summarized in Table 7.1. Six earlier cohort studies each involved less than 1000 subjects.

The two cohort studies of stomach cancer mortality in hospital case series produced conflicting results. The UK study (Caygill et al. 1986) involved almost 4500 patients, operated in one hospital during the period 1940–1960, who were followed up for at least 25 years. Gastric cancer mortality was significantly increased (relative risk 1.6), and the risk increased with the time elapsed since surgery, more so for gastric than for duodenal ulcer (Table 7.2). Minor differences in risk were seen for the different surgical procedures (Billroth I, Billroth II and vagotomy with gastroenterostomy). By contrast, the Japanese mortality study (Kuratsune et al. 1986), although similar in design, produced an overall relative risk of gastric cancer that was significantly low (0.34), declining further to 0.21 at 10 years or more after surgery. This result is unlikely to be due to systematic under-ascertainment of cancer deaths, since the numbers of deaths observed from all causes and all cancers were very close to the numbers expected in both sexes, and there were significant excesses of cancers of the liver and lung. The authors suggest that gastric cancer mortality may be reduced by intense surveillance of post-gastrectomy patients in Japan, where the background incidence of gastric cancer is very high, but no data are given in support of this argument on the number of incident cases in the cohort.

The study in western Norway (Viste et al. 1986) identified 87 new cancers of the stomach arising from 6 to 57 years (median 28 years) after surgery, a significant 2-fold increase in risk relative to the general population of the region. The risk in the first 10 years after surgery was not high, but it rose steadily in successive 5-year intervals to reach 7-fold at 40 or more years after surgery. There was no major difference in risk between the different surgical procedures used. Three more recent cohort studies from Nordic countries have all produced similar results (Arnthorsson et al. 1988; Lundegårdh et al. 1988; Møller and Toftgaard 1990). The overall risk of gastric cancer following gastric surgery is not significantly high; there is a low risk

Surgery

Table 7.1 Gastric cancer following partial gastrectomy for benign disease: cohort studies

Source	Country	Recruitment period[1]	End of study	Endpoint	Number of subjects	Observed	Expected	RR[2]	95 per cent CI[2]
Kuratsune et al. 1986	Japan	1948–1970	1981	Mortality	3827	34	100.63	0.34	0.2–0.5[3]
Caygill et al. 1986	UK	1940–1960	1985	Mortality	4466	80	50.7	1.58	1.25–1.96[3]
Viste et al. 1986	Norway	1900–1969	1984	Incidence	3470	87	41.4	2.1	1.7–2.6[3]
Arnthorsson et al. 1988	Iceland	1930–1974	1982	Incidence	1795	40	34.4	1.16	0.83–1.58
Lundegårdh et al. 1988	Sweden	1950–1958	1983	Incidence	6459	102	106.8	0.96	0.78–1.16
Møller and Toftgaard 1990	Denmark	1955–1960	1987	Incidence	4107	59	55.14	1.07	0.8–1.4[3]

[1] Period during which subjects underwent surgery; all the cohorts were assembled retrospectively.
[2] RR (relative risk) = obs/exp; 95 per cent CI = 95 per cent confidence interval.
[3] Confidence intervals estimated from published data.

Table 7.2 Gastric cancer following partial gastrectomy for benign disease: risk by time since operation

Source	Diagnosis	0–19 years				20+ years			
		Obs	Exp	RR	95 per cent CI	Obs	Exp	RR	95 per cent CI
Caygill et al. 1986	All types	44	42.3	1.04	0.8–1.4	36	8.4	4.39	3.0–5.9
	DU	9	20.9	0.43	0.2–0.8	20	5.4	2.70	2.3–5.6
	GU	32	12.0	2.67	1.8–3.8	12	2.2	5.45	3.0–9.3
Viste et al. 1986				1.4				3.9	
Arnthorsson et al. 1988		20	26.7	0.75	0.5–1.2	20	7.7	2.6	1.6–4.0
Lundegårdh et al. 1988		40	69.5	0.57	0.4–0.8	62	37.3	1.66	1.3–2.1
Møller and Toftgaard 1990	All types	36	42.4	0.85	0.6–1.2	23	12.8	1.80	1.1–2.7
	DU	15	13.9	1.08	0.6–1.8	7	3.4	2.06	0.8–4.2
	GU	17	25.4	0.67	0.4–1.1	14	8.3	1.68	0.9–2.8

Obs, observed; Exp, expected; RR, relative risk; CI, confidence interval; DU, duodenal ulcer; GU, gastric ulcer.

in the first 10–20 years after surgery, significantly low in two of the studies. A significantly high risk is observed in all of the studies 20 or more years after surgery (Table 7.2). The Swedish study (Lundegårdh et al. 1988), which is the largest, showed that the risk of stomach cancer increased on average 1.28-fold in each successive 5-year period after surgery, after adjustment for sex, age, diagnosis, and type of operation, each of which also affected the risk independently. The Billroth II procedure was associated with a higher risk than the Billroth I, and gastric ulcer with a 2-fold greater risk than duodenal ulcer. For each of four successive age groups at surgery (under 40, 40–49, 50–59, 60+), the overall risk decreased by about 50 per cent.

Among four case-control studies of gastric cancer following gastrectomy, two were autopsy series and two hospital case-control studies based on medical records. The autopsy series by Stalsberg and Taksdal (1971) included 630 cases of gastric cancer and 630 control autopsies. Gastric surgery for benign disease at least 5 years earlier was three times more frequent in cases than in controls, and risk rose steadily to more than 5-fold at 15 or more years since surgery. Bias is unlikely to explain this result. The study by Sandler et al. (1984) included 521 cases of gastric cancer and an equal number of hospital controls matched for age, sex, and year of admission. The overall odds ratio for gastric surgery at least 5 years before admission was 0.7 (95 per cent CI 0.3–1.6), but again, the risk increased with progressively longer intervals: for gastric surgery at least 10, 15, and 20 years earlier, the odds ratios were 1.8, 2.3, and 5.0, respectively.

The likely mechanisms for a carcinogenic effect of partial gastrectomy upon the remnant stomach include both hypochlorhydria, which is an objective of the operation, and reflux of bile acids and duodenal contents into the stomach, which is a complication of the drainage procedures used to replace the pyloric sphincter. If reflux of duodenal contents into the stomach is an important mechanism, the risk of gastric cancer should be lower after highly selective vagotomy, which leaves the pylorus intact, but there appear to be no data on this point, perhaps because this operation is relatively recent.

Viste et al. (1986) argued from the increased incidence of gastric cancer which they observed after partial gastrectomy to suggest that operations other than resection should be considered for peptic ulcer in young patients, and that active clinical surveillance should be offered to all gastrectomy patients 15 years after surgery, but doubts about the efficacy of such a programme have been expressed both in the UK (Logan and Langman 1983) and in the USA (Sandler et al. 1984).

Cancer at other sites

Increased risk of cancer at sites other than the stomach following partial gastrectomy was also reported from the British and Japanese cohort studies (Table 7.3). The excess of lung cancer mortality may reflect the

Table 7.3 Other cancers following partial gastrectomy for benign disease

Cancer site	Japan[1]		United Kingdom[2]				Denmark[3]			
			0–19 years		20+ years		0–19 years		20+ years	
Country:	Number of deaths	RR	Number of deaths	RR	Number of deaths	RR	Number of cases	RR	Number of cases	RR
Oesophagus	7	0.60	9	0.8	6	2.3	9	1.49	3	1.21
Intestine and colon	13	1.73 }	37	0.7*	20	12.6*	65	0.89	40	0.83
Rectum	15	1.70								
Liver	47	2.29**	–	–	–	–	2	0.44	3	1.05
Biliary tract	–	–	6	2.0	6	9.4***	8	1.34	1	0.39
Lung	41	1.44*	215	1.3*	146	3.9***	125	1.59***	77	1.80***
Pancreas	4	0.43	13	0.7	15	3.8*	19	1.05	13	1.44
Breast (female)	–	–	6	0.45*	9	4.0***	21	0.88	9	0.89
All sites	–	0.95	467	1.0	323	3.3***	601	1.17***	329	1.28***

* $p<0.05$; ** $p<0.01$; *** $p<0.001$.
[1] Kuratsune et al. 1986.
[2] Caygill et al. 1987a,b; 1988.
[3] Møller and Toftgaard 1990.

confounding influence of cigarette smoking, which is also associated with peptic ulceration, but the size and generality of the excess cancer mortality 20 years or more after gastric surgery in the British study is puzzling, and is difficult to accept without confirmation. In the recent Danish study (Møller and Toftgaard 1990), which attempted to replicate these findings, only the excess of gastric cancer and lung cancer was confirmed, although the risks were lower (Table 7.3). The mechanism by which partial gastrectomy might induce cancer in distant organs is unclear; Caygill et al. (1987a,b) have proposed that a circulating N-nitroso carcinogen may be produced by the gastric remnant after an interval of 20 years or more.

In summary, partial gastrectomy for benign disease undoubtedly increases the risk of subsequent cancer of the gastric remnant, probably as an indirect result of long-term reflux of duodenal contents. Any sizeable increase in cancer risk elsewhere as a result of the operation appears unlikely.

Cholecystectomy

Cholecystectomy for benign disease is usually carried out for symptomatic cholelithiasis, less often for acute cholecystitis. The circulation and metabolism of bile acids is substantially altered by the surgery, and there is some evidence to suggest that these changes might increase the risk of colon cancer. After cholecystectomy, bile is secreted continuously into the gut instead of intermittently in response to meals, the total bile salt pool is reduced, and the proportion of deoxycholic acid and other secondary bile acids in the bile acid pool increases, as a result of increased exposure of bile salts to hydroxylating enzymes from intestinal bacteria (Pomare and Heaton 1973). There was significantly increased faecal excretion of deoxycholic and lithocholic acids in a small study of patients with colon cancer and adenomatous polyps (Reddy and Wynder 1977), and these acids have been shown to promote colonic carcinogenesis induced by an N-nitroso carcinogen in the rat (Narisawa et al. 1974). Cholecystectomy in the mouse increased the risk of colon cancer by 4-fold (from 16 to 70 per cent) after administration of the carcinogen dimethylhydrazine, whereas there were no cancers in the group treated by cholecystectomy alone (Werner et al. 1977). These observations suggest that the metabolic changes resulting from cholecystectomy might promote carcinogenesis in the human colon.

The evidence from human studies of cholecystectomy as a risk factor for colon cancer is, however, difficult to interpret. Whilst the majority of published studies show an increased risk, some of these depend for evidence of previous operations on clinical records, where such information is frequently unavailable. Both hospital record studies, and, in particular, autopsy series may also be subject to selection bias. In a series of 706 colon cancers selected from the Third National Cancer Survey, for example, Vernick et al. (1980) found a marked predominance of previous cholecystectomy in patients with right-sided colon cancer compared to those with left-sided colon cancer, whereas in a later, unselected series of 582 cases, Abrams et al. (1983) found no difference in the sub-site distribution of colon cancer (ascending, transverse, etc.) between patients with and without a previous cholecystectomy.

Table 7.4 summarizes the results of some of the case-control studies. The largest of these, based on the members of a private health insurance plan, is negative, and its authors (Friedman et al. 1987) propose that the association between cholecystectomy and colon cancer observed in previous studies may be due to ascertainment bias, in which patients who have undergone cholecystectomy are subject to more intense observation and investigation of abdominal symptoms than other persons, resulting in diagnosis of more colon cancer. Such a bias, however, might be expected to increase the proportion of colon cancers diagnosed at an early stage in patients with a previous cholecystectomy, but Abrams et al. (1983) showed that the proportion of early cases (Dukes stage A) was similar among patients with and without a previous cholecystectomy. In a case-control study with no overlap with their previous case series, Vernick and Kuller (1981) compared right-sided colon cancers (cases) to left-sided colon cancers (controls), finding an odds ratio for cholecystectomy of 1.9. This result does not reach conventional significance, but the internal comparison is a more stringent test of variation in the strength of the association between previous cholecystectomy and cancer among different segments of the colon.

Overall, in the positive case-control studies, the odds ratios for cholecystectomy are consistently higher in women than in men, and consistently higher for the right colon than the left, even though the anatomical boundary used for this distinction varies slightly between the studies. Although such a pattern of evidence could still be the result of bias, the overall consistency is notable.

In one cohort study of 1681 persons who had a cholecystectomy in the period 1950–1969 and were followed up for a median of 13 years (Linos et al. 1981), the relative risk for cancers of the colon and rectum combined was 1.3 in both sexes. For colon cancer only, the relative risk was 1.7 in women and 0.9 in men. For right-sided colon cancer, the relative risk of 2.1 was significantly high in women (13 cases, 95 per cent CI 1.1–3.6), and high (1.7) but not significantly so in men (four cases, 95 per cent CI 0.5–4.3). This cohort study represents more or less complete follow-up of all persons undergoing cholecystectomy in a defined community and time period, and its results are consistent with those from the positive case-control studies. Its results are nevertheless contradicted by a larger cohort study; Adami

Table 7.4 Cancer following cholecystectomy for benign disease

Case–control studies	Country		Cancer site	Sex	Cases Chole[1]	Total	Controls Chole[1]	Total	Odds ratio	95% CI
Capron et al. 1978	France	Hospital records Autopsy controls	Colon	M F	3 14	129 108	18 32	1480 978	1.93 4.40	0.6–6.5[2] 2.4–8.1[2]
Vernick and Kuller 1981	USA	Hospital cases Telephone interview	Right colon (cases) Left colon (controls)		21	150	12	150	1.87	0.89–3.91
Turunen and Kivilaakso 1981	Finland	Autopsy records	Colon, rectum		45	304	32	304	1.59	1.01–2.55[3]
Mamianetti et al. 1983	Argentina	Hospital records	Colon, rectum		17	124	19	124	0.87	0.43–1.73[3]
Friedman et al. 1987	USA	Hospital records	Colon, rectum		174	5898	773	27687	1.1	0.9–1.2
Herrera Hernández et al. 1987	Mexico	Hospital records	Colon, rectum	M F	7 18	100 100	9 17	100 100	0.76 1.07	0.27–2.14 0.5–2.3
Cohort studies					Number of subjects	Observed cases	Expected cases		Relative risk	
Linos et al. 1981	USA	Hospital records 1950–1969	Colon, rectum	M F	460 1221	13 29	10.0 22.2		1.3 1.3	0.7–2.2 0.9–1.9
Adami et al. 1983, 1984	Sweden	Hospital records 1964–1967	Colon, rectum	M F	5095 11678	46 84	58.3 95.2		0.79 0.88	0.58–1.05 0.70–1.09
			Breast	F	11678	202	199.1		1.0	0.9–1.2[4]

[1] Chole, number of subjects who had had a cholecystectomy.
[2] Unmatched odds ratio and 95% confidence interval calculated from published data.
[3] 90% confidence intervals.
[4] 95% confidence interval calculated from published data.

et al. (1983) identified a cohort of 11 678 women who underwent cholecystectomy for benign disease in the period 1964–1967 in the Uppsala region of Sweden (population 1.3 million) and followed them up for 11–14 years. Incident cancers were identified by linkage to the national cancer registry, which is considered largely complete, and deaths were identified through the national death registry. No increase in colorectal cancer incidence was observed over that of the population from which the cohort was drawn (Table 7.4).

It is known from autopsy and other studies that not all gallstones cause symptoms, and not all patients with symptoms undergo cholecystectomy. It is possible, therefore, that the observed excess of cholecystectomy in colon cancer patients is simply due to the underlying cholelithiasis rather than the surgical treatment of it, or that cholelithiasis and colon cancer themselves share some common underlying aetiology such as diet, or obesity. Thus one matched case-control study of colon cancer with 109 case-control pairs gave an odds ratio of 2.4 (95 per cent CI 1.2–4.7) for current cholelithiasis, but no significant increase in risk for previous cholecystectomy (Gafà *et al.* 1987). In contrast, one autopsy study showed a significant association between prior cholecystectomy and colon cancer, but no difference between cases and controls in the prevalence of unoperated gallstones (Turunen and Kivilaakso 1981; see Table 7.4). Similarly, there was no association between colon cancer and the diagnosis of symptomatic gallstones in the population studied by Linos *et al.* (1982), in contrast to the association observed following the surgical procedure to remove them (Linos *et al.* 1981).

The paradox therefore arises that there is a plausible mechanism for carcinogenesis after cholecystectomy—the operation leads to increased exposure of the gut to secondary bile acids such as deoxycholic acid, which have carcinogenic activity in experimental animals and are excreted in greater quantity by patients with colon cancer than by other persons—yet the epidemiological evidence that this surgical procedure increases the risk of colon cancer is not entirely convincing. It is impossible to dismiss the evidence that cholecystectomy may increase the risk of colon cancer to some extent, however, and in view of the increasing frequency with which this operation is performed (over 300 000 a year in the USA in 1970) and the prevalence of cholecystectomy in the middle-aged population (up to 10 per cent), the public health importance of any such risk for colon cancer could be considerable.

Breast cancer risk has also been studied in the large Swedish cohort of women who had undergone cholecystectomy (Adami *et al.* 1984): there appears to be no excess risk (Table 7.4). The number of breast cancers observed in this study was almost identical to the number expected from breast cancer incidence rates in the general population of the region, and there was no significant increase in risk in any of three time intervals since cholecystectomy (0–4, 5–9, 10–14 years) or in any of five 10-year age groups up to 70 years and over.

Organs with immunological function

In view of the increased risk of lymphoma and other malignancies that is associated with immune deficiency states (Biemer 1990), whether induced by infection, for example with the human immunodeficiency virus, or deliberately, in order to avoid graft rejection in transplant patients (Dorreen and Hancock, Chapter 5, this volume), it is of some interest to know if cancer risk is altered in patients who have undergone surgical removal of organs which have an immunological function or which contain lymphoid tissue.

Splenectomy and thymectomy

Immunological functions of the spleen include both humoral and cellular responses to foreign antigens and the clearance of particulate antigens from the circulation (Haynes 1987), and splenectomy may result in inability to mount an adequate immune response to bacterial infection (Wolf and Neiman 1989). Splenectomy is generally carried out following traumatic rupture, or as part of the staging procedure for a lymphoma, or in the treatment of some of the haemolytic anaemias.

Cancer risk following splenectomy for trauma has been investigated in a cohort of 740 US servicemen operated upon during 1944–1945 and followed up until the end of 1974 (Robinette and Fraumeni 1977). Their mortality was compared with that of an equal number of servicemen admitted to hospital for nasopharyngitis in the same period and individually matched on race and year of birth, and with mortality in the general population. There was significant excess mortality in the splenectomized men relative to the comparison group for all causes (relative risk 1.46), for ischaemic heart disease ($RR = 1.9$), and from pneumonia, the excess arising largely 10 or more years after splenectomy, but cancer mortality was not increased ($RR = 0.67$, based on 14 deaths in the splenectomy group).

In contrast, splenectomy for staging of Hodgkin's disease significantly increased the risk of subsequent leukaemia and myelodysplastic syndrome ($RR = 3.61$, based on 22 cases) in a cohort of 744 patients treated in the Netherlands during the period 1966–1983 (van Leeuwen et al. 1987). No relation with splenectomy was observed for other second malignancies.

The thymus is the site of T-lymphocyte maturation in the fetal and early postnatal period; it involutes after puberty but remains functional in adult life. T-lymphocytes are responsible for cell-mediated immunity, and cytotoxic T-cells may lyse virally-infected cells and tumour cells (DeVita et al.

1989). Neonatal thymectomy in experimental animals increases the incidence of malignancy induced by viral and chemical agents (Kinlen 1982). Thymectomy is used in the treatment of myasthenia gravis and for tumours of the thymus, both rare.

The concept of immunological surveillance as a natural defence against cancer prompted the examination of cancer mortality among 381 patients who underwent thymectomy for myasthenia gravis in London during the period 1942–1964; most of the patients were in the age range 20–39 years, and 65 of them had a thymoma (Vessey et al. 1979). The patients were followed for up to 30 years, to the end of 1972. Two thirds of the 148 observed deaths were due to the underlying cause for surgery, and mortality from all other causes was slightly higher than expected on the basis of national mortality rates (42 deaths observed versus 29.8 expected, $p = 0.03$), but mortality from tumours other than those of the thymus was not significantly high (11 versus 8.78). A further 11 incident tumours were observed. Mortality from extra-thymic tumours did not appear to be associated with age at thymectomy or the time interval since the operation, and there was no difference between patients with and without a thymoma.

Appendectomy and tonsillectomy

Appendectomy for acute inflammation of the vermiform appendix has been widely practised for over 100 years. It is usually a minor operation, providing definitive treatment for a condition which may otherwise progress to abscess or peritonitis. It is considered to be so free of risk that a normal appendix is often removed incidentally at laparotomy for other conditions, in order to prevent subsequent appendicitis. The appendix is usually considered to be a purely vestigial organ: it does contain lymphoid tissue, but whether this has any unique function, either at a particular age or throughout life, is unknown. The tonsils also contain lymphoid tissue, but whether tonsillectomy has any lasting effect on immune function appears to be unknown, although impairment for up to 7 months of the nasopharyngeal secretion of antibodies to polio virus has been reported following tonsillectomy (Ogra 1971).

The evidence on appendectomy and tonsillectomy as risk factors for cancer has been reviewed by Lee (1975) and Howson (1983). The first study to suggest that there might be adverse late effects of appendectomy was a careful analysis of over 900 autopsies in Kansas, USA, reported by McVay (1964), who found a higher prevalence of appendectomy in autopsied patients dying of cancer than in a control group dying of cardiovascular disease; the excess prevalence (19 per cent versus 8 per cent) was significant for colon cancer (227 deaths), but was also present for cancers of the lung, breast, cervix, stomach, and pancreas. Ten case-control studies were reviewed by Howson (1983), most of them showing a positive association between prior appendectomy and, variously, Hodgkin's disease and cancers of the breast, colon and pancreas. Several of these early studies failed to control for social class or even for age, however, and a number used hospital medical records alone to obtain information on previous operations.

Two well-designed case-control studies of cancer of the pancreas have been reported in which exposure was defined as appendectomy at least ten years (Howson and Asal 1975) or five years (Haines et al. 1982) previously, and information was obtained at interview. The first study was positive, with an odds ratio of 2.1 (95 per cent CI 1.2–3.7, based on 201 cases), while the second was negative, with an odds ratio of 0.9 (95 per cent CI 0.5–1.6, based on 116 cases). A large case-control study of colorectal cancer (Vobecky et al. 1983) showed a significant association with appendectomy for colon cancer in men but not in women, and no effect for rectal cancer.

The risk of Hodgkin's disease has been reported as high both after appendectomy and after tonsillectomy. Hyams and Wynder (1968) reported a significant excess of previous appendectomy in men from their data, with an odds ratio of 3.6 (95 per cent CI 2.0–6.3, based on 100 cases), although there was no significant difference for women. In the case-control study by Vianna et al. (1971a), including 109 matched pairs, the odds ratio given for tonsillectomy was 2.9.

The only cohort study of appendectomy and cancer (Moertel et al. 1974) involved 1779 persons in Rochester, Minnesota, who underwent appendectomy in the period 1925–1944, and 1943 persons living in the same town, matched by age, sex, and period of observation, who had had dental extractions or fractures in the same period, but had never had an appendectomy. Both groups were followed up for a mean of 21 years to the end of 1971, and 288 non-skin malignancies were observed. The ratio of the number of malignancies observed to the number expected on the basis of Connecticut cancer incidence rates (relative risk) was similar in the appendectomy and comparison groups (0.88 versus 0.85). Furthermore, the relative risks were similar in groups variously defined as having had neither appendectomy nor tonsillectomy, either operation alone, and both. The reason for appendectomy was acute or chronic appendicitis in 39 and 35 per cent respectively, but in 26 per cent the appendix was removed incidentally at surgery carried out for other reasons. In view of the possible role of the appendix in developing immunity to intestinal viruses, it would have been interesting to know if cancer risk differed between those groups.

The pattern of human evidence linking appendectomy and tonsillectomy with subsequent cancer risk remains inconclusive, particularly in view of the limited understanding of the normal functions (if any) of these tissues, and whether their removal, and the age at removal, has any lasting effect on immune competence. Involution of the tonsillar lymphoid tissue in early adolescence may be important, particularly for subsequent Hodgkin's disease. The epidemiological evidence is intriguing (Vianna et al. 1971b), and it remains consistent with McVay's (1964) advice to surgeons that 'the present practice of removing a normal appendix during surgery for other reasons should be reconsidered'. There is still a need for a well-designed study of subjects who have undergone appendectomy or tonsillectomy, in which the age at operation, the indications for surgery and the time since it was carried out are taken into account in the analysis of subsequent cancer risk.

Vasectomy

Several studies of the relationship between vasectomy and subsequent cancer risk are available, principally of cancers of the testis and prostate. Only one offers any real support for an association with prostate cancer. In a matched-pair case-control study of prostate cancer in men aged less than 60 in California, Honda et al. (1988) reported an odds ratio for vasectomy in ever-married men of 1.4 (95 per cent CI 0.9–2.3). There was a highly significant upward trend in the risk of prostate cancer with time elapsed since vasectomy, and after 30 or more years the risk was four times greater than in men who had not had a vasectomy. Although cases were identified from a population-based cancer registry, however, only 216 (55 per cent) of eligible cases could be interviewed, and risk factor information was obtained by telephone interview, both factors which might bias the risk upward in a study of this nature.

Testosterone is involved in the growth and development of the prostate, and is likely to be involved in the aetiology of prostate cancer. A measure of free circulating testosterone was 13 per cent higher ($p<0.03$) in vasectomized than non-vasectomized men in a small sample of controls in this study, but it is unknown whether this difference was also present before surgery, and thus whether it was an effect of surgery which might help to explain the association observed between vasectomy and prostate cancer risk.

In a hospital-based case-control study designed to examine many possible associations between exposure and disease, Rosenberg et al. (1989) found a history of vasectomy in 10 per cent of prostate cancer cases. The odds ratio for a previous vasectomy in prostate cancer was 3.0 ($p<0.05$), and was unaffected by adjustment for cigarette smoking.

The incidence of prostate cancer has been studied in a group of over 5000 men who reported having had a vasectomy at some time in the past when completing a questionnaire at entry into a private health insurance scheme in the period 1977–1982. These men were compared to a group of 15 000 men, frequency-matched on age, race and marital status, who did not report a prior vasectomy at entry into the scheme during the same period (Sidney 1987). The men were followed up for an average of 4.6

years to the end of 1984. The incidence of prostate cancer was similar in the two groups (relative risk 1.0, 95 per cent CI 0.6–1.7, based on 68 cases with a prior vasectomy), and the relative risk was similar in men who had had a vasectomy up to 20 years and 20 or more years previously. Although this study is negative, the period of observation was short, and there is some question as to whether ascertainment of cases was complete. A study of similar design (Petitti *et al.* 1983), involving over 4000 vasectomized men and 13 000 men without a prior vasectomy, also failed to detect an increase in overall cancer risk.

It seems unlikely that vasectomy increases the risk of testicular cancer. The odds ratio for a prior vasectomy was 0.6 (95 per cent CI 0.3–1.2) in one case-control study of testicular cancer among men aged 18–40 years (Moss *et al.* 1986). In another case-control study in men aged 20–69 years (Strader *et al.* 1988), the odds ratio for vasectomy was 1.5 (95 per cent CI 1.0–2.2), but the excess risk was entirely restricted to Catholic men, among whom the odds ratio was 8.7 (95 per cent CI 2.8–27.1). Again, after various exclusions, only about half of the cases identified were included in this analysis, and information on vasectomy was obtained by telephone interview. Bias, including under-reporting of vasectomy by Catholic controls, is likely to be responsible for this striking result. Two large cohorts of vasectomized men have reported no increase in overall cancer risk (Petitti *et al.* 1983; Massey *et al.* 1984).

Surgical implantation of a foreign body

Surgical implantation of a foreign body designed to remain in place indefinitely is now a common procedure. Implants include cardiac pacemakers, joint prostheses, aneurysm clips, and breast prostheses. Case reports of cancer in individuals with such implants are frequent, but these are only of limited use in assessing whether there is any causal relationship, and virtually none for estimating the magnitude of any associated risk. Adequate human studies are rare. Animal studies have provided a good deal of information on the mechanisms of carcinogenesis associated with foreign bodies, but there is considerable variation between animal species in susceptibility, and the risks observed in animals cannot necessarily be inferred in humans.

Penn (1982) has pointed out that most tumours induced by foreign bodies are sarcomas, and that physical features of the implant appear more important in carcinogenesis than its chemical composition: hard, smooth, hydrophobic materials are more carcinogenic in animals than soft, rough and hydrophilic materials. These properties of the implant influence the degree of fibrosis which it provokes. The extent of fibrosis in response to a foreign body varies between animal species, and this may account for some of the

Breast implants

Surgical implants to increase the size of the female breast have become widespread in developed countries since the introduction of silicone gel-filled prostheses in the early 1960s. In the USA, about 130 000 such operations are performed every year, mostly for cosmetic reasons, and an estimated two million women in the USA currently have such implants, which represents about 3 per cent of the adult female population. The US Food and Drug Administration recently cited an experiment, carried out by a manufacturer of silicone gel, which showed an excess of sarcomas in silicone-implanted rats, but pointed out that humans are less susceptible to the induction of sarcoma from subcutaneous implants than many animals, including the rat, adding that 'a carcinogenic effect in humans could not be ruled out, but that if such an effect did exist, the risk would be very low' (Anon. 1989).

The only large cohort study of cancer risk in women who had undergone augmentation mammoplasty was reported by Deapen et al. (1986). They identified 3500 women who had undergone this procedure in the period 1960–1979 at one of 35 large private plastic surgery clinics in Los Angeles, and followed up 3111 (89 per cent) of them to the end of 1981, a median of 6.3 years after the operation. Cancers were detected by careful matching of the cohort against the Los Angeles County Cancer Surveillance Program, a population-based cancer registry. Nine breast cancers were observed versus 15.7 expected, a relative risk of 0.57 (95 per cent CI 0.26–1.09). For all other cancers combined, however, there was a 50 per cent excess: 24 observed versus 15.8 expected, a relative risk of 1.52 (95 per cent CI 0.98–2.27). Overall cancer risk in this cohort was thus no different from that expected (33 versus 31.5).

More than 80 per cent of the women were aged less than 40 years at operation, and in these women only one breast cancer was observed (8.6 expected), giving a significantly low relative risk of 0.12 (95 per cent CI 0.03–0.65), whereas among the 18 per cent of the women aged 40 or over at operation, eight breast cancers were observed (7.1 expected). Relatively few of the women (447, 14 per cent) were followed up for 10 years or more, and among these women there were three breast cancers (3.5 expected). The study does not therefore offer much information on the long-term risk of breast cancer after augmentation mammoplasty, but follow-up is being continued to address this question, which is likely to become increasingly important as the pool of young women who have had the operation increases.

Two other large studies of women with augmentation mammoplasty have been reported, both done by surveying plastic surgeons. Harris (1961) sent questionnaires to all certified plastic surgeons in the USA: 184 surgeons provided data on some 16 660 women with implants, but not one confirmed malignant neoplasm of the breast was reported. De Cholnoky (1970) obtained data from 265 surgeons around the world, reporting one breast cancer and eight other cancers among 10 941 young women with breast implants followed for an average of about 3 years. Both studies are potentially biased toward under-ascertainment of cases, and neither can provide estimates of the number of cancers expected, in order to permit evaluation of the risk in relation to women without breast implants. These large studies are practically useless as evidence for the absence of any excess cancer risk associated with augmentation mammoplasty.

If silicone implants appear unlikely to increase the risk of breast cancer, at least in the first 10 years, the question has been raised of whether they may delay clinical or mammographic diagnosis of those tumours that do develop (Roberts and Taylor 1990). Silverstein *et al.* (1988) reported on a case series of 753 women with breast cancer seen at a centre specializing in breast disease during the period 1981–1986. Twenty (2.7 per cent) of the women had undergone augmentation mammoplasty up to 15 years (mean 6.9 years) earlier. The proportion of this small group with positive axillary lymph nodes was significantly higher than in the remaining women (65 per cent versus 28 per cent), and none of them was diagnosed by mammography, compared to 21 per cent of the remaining women. The authors claimed that silicone implants reduce the ability of mammography to detect breast tumours, but presented no evidence as to whether the 20 women with breast implants had ever had a mammogram while still asymptomatic: the lesions were nevertheless visible on mammograms in 10 of the 15 women who underwent this procedure prior to biopsy. Miller (1989) has also pointed out that modified mammographic views permit full examination of the augmented breast.

Although surgical *reduction* of the hypertrophic female breast does not involve implants, it is relevant to discussion of breast cancer in relation to the mass of breast tissue at risk. Lund *et al.* (1987) identified 1245 women who had a reduction mammoplasty in one of five Copenhagen hospitals during the period 1943–1971, and followed them up to the end of 1982 by linkage to the Danish Cancer Registry. Over 60 per cent of the women were aged less than 40 years at operation. Eighteen breast cancers were observed, versus 30.28 expected on the basis of national cancer incidence rates, giving a significantly low relative risk of 0.59 (95 per cent CI 0.35–0.94). When examined by the time elapsed since surgery, the reduction in risk improved from 30 per cent in the first 10 years after operation to 44 per cent after 10 or more years. This result cannot be satisfactorily explained on the basis that reduction mammoplasty involves unwitting removal of occult breast cancers, since this would be expected to produce an early reduction in risk followed by a rise toward the normal or underlying pattern, exactly the opposite of what was observed. Furthermore, the risk reduction was directly related to the amount of breast tissue removed: 16 per cent for less than 400 grams per breast, 34 per cent for 400–599 grams, and 69 per cent for 600 grams or more.

The few available studies of surgical manipulation of the female breast do not therefore suggest any increase in the risk of subsequent cancer of the breast, but they do offer an insight into carcinogenesis. Removal of an organ obviously eliminates the risk of new cancers arising from it, but it is not so obvious that cancer risk in an organ is proportional to its size, and thus perhaps to the number of cells at risk of malignant transformation which it contains; yet the available evidence on breast cancer following plastic surgery is consistent with this idea. Deapen *et al.* (1986) speculate that the low breast cancer risk after agumentation mammoplasty in women aged less than 40 years at operation may relate to a smaller initial volume or cell mass of susceptible breast tissue, whereas the older women, among whom there was no reduction in risk, might have sought the operation after atrophy of breasts of more normal size.

Whether surgery to increase breast size with silicone gel prostheses may increase the risk of cancer in organs other than the breast remains unclear. In the only relevant study, that of Deapen *et al.* (1986), the relative risk of 1.52 (95 per cent CI 0.98–2.27) is suggestive of an increase, and the results of further follow-up of this cohort will be of considerable interest.

Orthopaedic prostheses

Man-made materials are often implanted temporarily to provide physical support for the repair of complex fractures, and permanent replacement of joints irreparably damaged by fracture or arthritis, particularly the hip joint, has become progressively more common over the last 50 years. The materials used have usually been metal alloys containing variously cobalt, chromium, nickel, and molybdenum, chosen for their strength, lightness, and resistance to corrosion. More recently, plastics and organic cements have been used as well.

Prolonged contact of metal alloys with body fluids results in gradual corrosion of even the most inert metals, and increased serum, urine, and tissue concentrations of cobalt, chromium, and nickel have been reported in patients with alloy prostheses containing these metals (Sunderman 1989). Corrosion of the metal components of a typical total hip replacement has been estimated to release up to 20 mg of metal a year, while the total body burden of such metals in a 70 kg man is of the order of 10 mg (Black 1988). Cobalt metal powder and cobaltous oxide are carcinogenic in experimental animals, and there is some evidence that metal alloys containing cobalt, chromium, and molybdenum may also be carcinogenic (IARC 1991).

Sarcomas have been reported at the site of metallic foreign bodies resulting from war injuries, and there are veterinary reports of sarcomas at the site of metal fixation implants in animals (Stevenson *et al.* 1982). There have also been 16 case reports of malignant neoplasia, mostly sarcomas, arising at the site of orthopaedic prostheses in humans (Sunderman 1989; IARC 1991). In most of these case reports, the prosthesis was made of a cobalt-containing alloy, often a cobalt–chromium–molybdenum alloy of the type known as Vitallium.

Only one cohort study of persons with a hip prosthesis has been reported (Gillespie *et al.* 1988). In this study, 1358 persons who underwent a total hip replacement in New Zealand in the period 1966–1973 were followed up for periods ranging from 6 months to 17 years (mean 10.6 years) until the end of 1983. New cancers were identified through the New Zealand cancer registry, and cancer incidence in the study cohort was compared to that in the general population. Overall, there were 164 cancers observed, not significantly different from the number expected (179.4 cases), giving a standardized morbidity ratio (SMR) of 91 (95 per cent CI 78–107). There was a significant excess of tumours of the lymphatic and haemopoietic system, however: (SMR 168, 95 per cent CI 106–260, based on 21 cases). Conversely, there were significant deficits of cancers of the breast (SMR 36, 95 per cent CI 14–82, based on six cases) and colon and rectum combined (SMR 41, 95 per cent CI 39–96, based on 21 cases). When results were examined by time since insertion of the hip prosthesis, the risk of all cancers combined was significantly low in the first 10 years after surgery (SMR 74, 95 per cent CI 61–90, based on 107 cases), but this deficit was no longer statistically significant if the 298 persons whose status at the end of the study could not be satisfactorily established were removed from the analysis. Furthermore, overall cancer risk was significantly high at 10 or more years since surgery (SMR 160, 95 per cent CI 122–209, based on 57 cases).

The results of this relatively small but well-designed study suggest that there is no cause for alarm about cancer risk in patients with metal alloy joint prostheses, especially in view of the obvious benefits of a successful joint replacement. The cumulative risk of developing a lymphoreticular neoplasm in the New Zealand study was still small, increasing from two per 1000 over 10 years in the general population to about six per 1000 in the population with a hip replacement.

In view, however, of the known carcinogenicity in animals of some of the component metals and of their compounds, of the increasing population of elderly people living for extended periods with a prosthesis, and in particular of the significantly increased cancer risk observed in the New Zealand study ten or more years after insertion of the prosthesis, these results are certainly not cause for complacency. The suggestion (Hamblen and Carter 1984) that further case reports of malignancy arising as a late complication of arthroplasty should be published 'so that any possible risk can eventually be quantified' is also inappropriate, since case reports are of little use for this purpose.

On the contrary, there is a clear need for a cohort study of patients with joint prostheses that is larger than the New Zealand study, covers a longer period of time, and includes if possible information on the type and composition of the alloys used, as well as on potential confounding exposures for the major cancers. Interpretation of the results would be greatly enhanced by measurements of the degree of corrosion of the implants and of concentrations of the relevant metal ions in body tissues and fluids, both for subjects who develop a cancer and for at least a sample of those who do not.

Cardiac pacemakers

Subcutaneous implantation of cardiac pacemakers is relatively recent, being common only since about 1970. Several cases of cancer arising in an organ near the pacemaker site have been reported, particularly of breast cancer in women. Dalal et al. (1980) reported two adenocarcinomas of the breast arising two and three years after insertion of a pacemaker in elderly women. They also calculated the number of breast cancers to be expected in the 333 women still under supervision in a university hospital cardiology clinic after insertion of a permanent pacemaker during the decade 1971–1979, obtaining a value of 1.85 expected breast cancers in 1040 woman-years at risk, similar to the number observed. This observation is more useful than a simple case report, representing a small retrospective cohort study prompted by the two cases actually observed, but it is limited to a single cancer type in women, and to the group of patients still being followed up, and is clearly subject to potential downward bias of the risk.

Many large cardiology centres should by now have access to large series of patients with permanent subcutaneous pacemakers, providing an opportunity for an adequately large and unbiased cohort study to be carried out rapidly, in order to assess any associated cancer risk, and including all cancers, both sexes and complete follow-up of all patients.

Cystoplasty and ureterosigmoidostomy

The use of segments of bowel to reconstruct the urinary tract has been common for 40 years. Augmentation cystoplasty is usually done to increase the size of a urinary bladder affected by tuberculosis, radiation or congenital anomaly. It involves prolonged exposure of the transplanted bowel segment to urine, with associated histological changes, as well as exposure

Surgery

of the adjacent bladder to bowel secretions. Fourteen cases of malignant neoplasm in an augmented bladder have been reported at intervals of 3 to 24 years after surgery, but Golomb *et al.* (1989) note the rarity (two cases) of this complication of augmentation cystoplasty in their series of over 400 patients operated between 1960 and 1987.

Ureterosigmoidostomy is a procedure for diversion of the urinary system away from the bladder, involving connection of the ureters to the sigmoid colon. It has been done mostly for congenital exstrophy of the bladder in infants, but also as an accessory procedure in adults following surgical removal of the bladder, or for intractable radiation cystitis. It was first done in 1852, but has to some extent been replaced by the use of ileal conduits since 1950. Reports of adenocarcinoma of the colon following ureterosigmoidostomy are numerous (e.g. Urdaneta *et al.* 1966; Lasser and Acosta 1975; Sooriyaarachchi *et al.* 1977; Parsons *et al.* 1977; Sheldon *et al.* 1983). There seems little doubt that the risk of adenocarcinoma of the colon is increased by ureterosigmoidostomy, partly because the tumours often arise in adults aged less than 40 years, but also because they almost invariably occur at the site of the anastomosis, in some cases even after the urine flow has been diverted elsewhere by further surgery. There is also a reasonable model of the sequence of events in rats, suggesting that both the urine flow and a faecal stream at the site of the anastomosis are necessary to initiate the development of the tumour (Crissey *et al.* 1980).

The risk of adenocarcinoma of the colon following ureterosigmoidostomy has, however, been grossly overestimated. Two incorrect estimates of the risk relative to the general population have been repeatedly cited—sometimes wrongly—by later authors. The first of these reports (Urdaneta *et al.* 1966) uses three cases (only two confirmed) of adenocarcinoma of the colon among 23 10-year survivors of ureterosigmoidostomy to derive a 'rate' (in fact a proportion) of 13 300 per 100 000 (3/23 = 13 043 per 100 000), which is then compared to an incidence rate of 24 per 100 000 per year in the general population under 45 years. In the second report (Parsons *et al.* 1977), a similar calculation is used to obtain a 'rate' of 6900 per 100 000 from two cases among 29 10-year survivors. The range of relative risks is then estimated to be 280-fold to 550-fold (6900/24 and 13 300/24). Leadbetter *et al.* (1979) combined their four cases in 90 10-year survivors with those from the first report to estimate a 5 per cent risk of colon cancer within 50 years of ureterosigmoidostomy, while the incorrect incidence rate of 6900 in the second report is later misquoted (Crissey *et al.* 1980; Sheldon *et al.* 1983) as a *relative* risk of 7000-fold.

A more plausible estimate of the incidence of adenocarcinoma of the colon following ureterosigmoidostomy in the series of Urdaneta *et al.* (1966) can be crudely derived from the data in their report, which covered a total of 83 patients (54 of them being treated for cancer) who underwent the procedure. This would be 463 per 100 000 person-years (three cases in

647 person-years).[1] This is still clearly higher than in the general population, perhaps 10- to 20-fold, but a more precise value cannot be calculated. It is worth noting, however, that in a similar but larger series of 160 patients with ureterosigmoidostomy followed for at least 10 years, reported in the discussion of this paper, no cancer of the bowel was observed.

It thus appears that none of the available studies provides an acceptable estimate of the risk, relative to the general population, of adenocarcinoma of the colon following ureterosigmoidostomy, although it is likely to be 10 to 30 times lower than is commonly stated. A retrospective cohort study of a large series of patients who have had ureterosigmoidostomy would enable the true risk, and its evolution with time since surgery, to be evaluated more precisely, and might also provide a useful new insight into bowel carcinogenesis, as well as a more rational basis for the clinical surveillance of patients who have a ureterosigmoidostomy for the development of subsequent malignancy.

[1] Total follow-up time estimated as follows: 56 persons died within 10 years, and four were lost to follow-up, estimate mean five years each (300 person-years), while the 23 ten-year survivors were followed for a mean of 15.08 years (347, total 647 person-years).

References

Abrams, J. S., Anton, J. R., and Dreyfuss, D. C. (1983). The absence of a relationship between cholecystectomy and the subsequent occurrence of cancer of the proximal colon. *Diseases of the Colon and Rectum*, **26**, 141–4.

Adami, H.-O., Meirik, O., Gustavsson, S., Nyrén, O., and Krusemo, U.-B. (1983). Colorectal cancer after cholecystectomy: absence of risk increase within 11–14 years. *Gastroenterology*, **85**, 859–65.

Adami, H.-O., Meirik, O., Gustavsson, S., Nyrén, O., and Krusemo, U.-B. (1984). Cholecystectomy and the incidence of breast cancer: a cohort study. *British Journal of Cancer*, **49**, 235–9.

Anon. (1989). Safety of silicone breast prostheses. *FDA Drug Bulletin*, **19**, 2–3.

Arnthorsson, G., Tulinius, H., Egilsson, V., Sigvaldason, H., Magnusson, B., and Thorarinsson, H. (1988). Gastric cancer after gastrectomy. *International Journal of Cancer*, **42**, 365–7.

Biemer, J. J. (1990). Malignant lymphomas associated with immunodeficiency states. *Annals of Clinical and Laboratory Science*, **20**, 175–91.

Black, J. (1988). Does corrosion matter? *Journal of Bone and Joint Surgery (Br)*, **70B**, 517–20.

Brand, K. G. (1975). Foreign body induced sarcomas. In *Cancer: a comprehensive treatise*, Vol. 1, *Etiology: physical and chemical carcinogenesis*, (ed. F. F. Becker), pp. 485–511. Plenum Press, New York.

Capron, J.-P., Delamarre, J., Canarelli, J.-P., Brousse, N., and Dupas, J.-L. (1978). La cholécystectomie favorise-t-elle l'apparition du cancer rectocolique? *Gastroentérologie Clinique et Biologique*, **2**, 383–9.

Caygill, C. P. J., Hill, M. J., Kirkham, J. S., and Northfield, T. C. (1986). Mortality from gastric cancer following gastric surgery for peptic ulcer. *Lancet*, **i**, 929–31.

Caygill, C. P. J., Leach, S. A., Kirkham, J. S., Northfield, T. C., Hall, C. N., and Hill, M. J. (1987*a*). Gastric hypoacidity as a risk factor for gastric and other cancers. In *Relevance of N-nitroso compounds to human cancer: exposures and mechanism*, (ed. H. Bartsch, I. K. O'Neill, and R. Schulte-Hermann), IARC Scientific Publications No. 84, pp. 524–6. International Agency for Research on Cancer, Lyon.

Caygill, C. P. J., Hill, M. J., Hall, C. N., Kirkham, J. S., and Northfield, T. C. (1987*b*). Increased risk of cancer at multiple sites after gastric surgery for peptic ulcer. *Gut*, **28**, 924–8.

Caygill, C., Hill, M., Kirkham, J., and Northfield, T. C. (1988). Increased risk of biliary tract cancer following gastric surgery. *British Journal of Cancer*, **57**, 434–6.

Crissey, M. M., Steele, G. D., and Gittes, R. F. (1980). Rat model for carcinogenesis in ureterosigmoidostomy. *Science*, **207**, 1079–80.

Dalal, J. J., Winterbottam, T., West, R. R., and Henderson, A. H. (1980). Implanted pacemakers and breast cancer. *Lancet*, **ii**, 311.

de Cholnoky, T. (1970). Augmentation mammoplasty: survey of complications in 10,941 patients by 265 surgeons. *Plastic and Reconstructive Surgery*, **45**, 573–7.

Deapen, D. M., Pike, M. C., Casagrande, J. T., and Brody, G. S. (1986). The relationship between breast cancer and augmentation mammaplasty: an epidemiologic study. *Plastic and Reconstructive Surgery*, **77**, 361–7.

DeVita, J. T., Jaffe, E. S., Mauch, P., and Longo, D. L. (1989). Hodgkin's disease. In *Cancer: principles and practice of oncology*, (ed. V. T. DeVita, Jr, S. Hellman, and S. A. Rosenberg) 3rd edn, pp. 1695–1740. Lippincott, Philadelphia.

Friedman, G. D., Goldhaber, M. K., and Quesenberry, C. P. (1987). Cholecystectomy and large bowel cancer. *Lancet*, **i**, 906–8.

Gafà, M., *et al.* (1987). Gallstones and risk of colonic cancer: a matched case-control study. *International Surgery*, **72**, 20–4.

Gillespie, W. J., Frampton, C. M. A., Henderson, R. J., and Ryan, P. M. (1988). The incidence of cancer following total hip replacement. *Journal of Bone and Joint Surgery (Br)*, **70B**, 539–42.

Golomb, J., Klutke, C. G., Lewin, K. J., Goodwin, W. E., deKernion, J. B., and Raz, S. (1989). Bladder neoplasms associated with augmentation cystoplasty: report of 2 cases and literature review. *Journal of Urology*, **142**, 377–80.

Haines, A. P., Moss, A. R., Whittemore, A., and Quivey, J. (1982). A case-control study of pancreatic carcinoma. *Journal of Cancer Research and Clinical Oncology*, **103**, 93–7.

Hamblen, D. L. and Carter, R. L. (1984). Sarcoma and joint replacement. *Journal of Bone and Joint Surgery (Br)*, **66B**, 625–7.

Harris, H. I. (1961). Survey of breast implants from the point of view of carcinogenesis. *Plastic and Reconstructive Surgery*, **28**, 81–3.

Haynes, B. F. (1987). Enlargement of lymph nodes and spleen. In *Principles of internal medicine*, (ed. E. Braunwald, K. J. Isselbacher, R. G. Petersdorf, J. D. Wilson, J. B. Martin, and A. S. Fauci), pp. 272–7. McGraw Hill, New York.

Herrera Hernández, M. F., Quintanilla Martínez, L., Viniegra Velázquez, L., and

Ponce de León Rosales, S. (1987). Cholecystectomy as a risk factor for the development of large bowel cancer. *Revista de Investigación Clínica (Méx.)* **39**, 101–6.

Hill, G. J. (1979). Historic milestones in cancer surgery. *Seminars in Oncology*, **6**, 409–27.

Honda, G. D., Bernstein, L., Ross, R. K., Greenland, S., Gerkins, V., and Henderson, B. E. (1988). Vasectomy, cigarette smoking, and age at first sexual intercourse as risk factors for prostate cancer in middle-aged men. *British Journal of Cancer*, **57**, 326–31.

Howson, C. P. (1983). Appendectomy and subsequent cancer risk. *Journal of Chronic Diseases*, **36**, 391–6.

Howson, C. P. and Asal, N. R. (1975). *An epidemiologic case-control study of pancreatic carcinoma*. University of Oklahoma Press, Norman.

Hyams, L. and Wynder, E. L. (1968). Appendectomy and cancer risk: an epidemiological evaluation. *Journal of Chronic Diseases*, **21**, 391–415.

IARC (International Agency for Research on Cancer) (1991). *IARC Monographs on the evaluation of carcinogenic risks to humans*, Vol. 52, *Chlorinated drinking water, chlorination by products, some other halogenated compounds; and cobalt and cobalt compounds*. IARC, Lyon (in press).

Kinlen, L. J. (1982). Immunologic factors. In *Cancer epidemiology and prevention*, (ed. D. Schottenfeld and J. F. Fraumeni, Jr), pp. 494–505. Saunders, Philadelphia.

Kuratsune, M., Inokuchi, K., Kumashiro, R., and Tokudome, S. (1986). Mortality from cancer after partial gastrectomy in Japan. *GANN Monograph on Cancer Research*, **31**, 213–19.

Lasser, A. and Acosta, A. E. (1975). Colonic neoplasms complicating ureterosigmoidostomy. *Cancer*, **35**, 1218–22.

Leadbetter, G. W., Zickerman, P., and Pierce, E. (1979). Ureterosigmoidostomy and carcinoma of the colon. *Journal of Urology*, **121**, 732–5.

Lee, Y.-T. N. (1975). Appendectomy, tonsillectomy and neoplasia. *Journal of Surgical Oncology*, **7**, 237–41.

Linos, D. A., Beard, C. M., O'Fallon, W. M., Dockerty, M. B., Beart, R. W., and Kurland, L. T. (1981). Cholecystectomy and carcinoma of the colon. *Lancet*, **ii**, 379–81.

Linos, D. A., O'Fallon, W. M., Thistle, J. L., and Kurland, L. T. (1982). Cholelithiasis and carcinoma of the colon. *Cancer*, **50**, 1015–19.

Logan, R. F. A. and Langman, M. J. S. (1983). Screening for gastric cancer after gastric surgery. *Lancet*, **ii**, 667–70.

Lund, K., Ewertz, M., and Schou, G. (1987). Breast cancer incidence subsequent to surgical reduction of the female breast. *Scandinavian Journal of Plastic and Reconstructive Surgery and Hand Surgery*, **21**, 209–12.

Lundegårdh, G., Adami, H.-O., Helmick, C., Zack, M., and Meirik, O. (1988). Stomach cancer after partial gastrectomy for benign ulcer disease. *New England Journal of Medicine*, **319**, 195–200.

Mamianetti, A., Cinto, R. O., and Lafont, D. (1983). Colecistectomia y adenocarcinoma colorectal. *Acta Gastroenterologica Latinoamericana*, **13**, 704–8.

Massey, F. J., et al. (1984). Vasectomy and health. Results from a large cohort study. *Journal of the American Medical Association*, **252**, 1023–9.

McVay, J. R. (1964). The appendix in relation to neoplastic disease. *Cancer*, **17**, 929–37.
Miller, S. H. (1989). Cancer surveillance after augmentation mammoplasty. *Archives of Surgery*, **124**, 134.
Moertel, C. G., Nobrega, F. T., Elveback, L. R., and Wentz, J. R. (1974). A prospective study of appendectomy and predisposition to cancer. *Surgery, Gynecology and Obstetrics*, **138**, 549–53.
Møller, H. and Toftgaard, C. (1990). Cancer occurrence in a cohort of patients surgically treated for peptic ulcer. *Gut* (in press).
Moss, A. R., Osmond, D., Bacchetti, P., Torti, F. M., and Gurgin, V. (1986). Hormonal risk factors in testicular cancer: a case-control study. *American Journal of Epidemiology*, **124**, 39–52.
Narisawa, T., Magadia, N. E., Weisburger, J. H., and Wynder, E. L. (1974). Promoting effect of bile acids on colon carcinogenesis after intrarectal instillation of *N*-methyl-*N'*-nitro-*N*-nitroso-guanidine in rats. *Journal of the National Cancer Institute*, **53**, 1093–7.
Ogra, P. L. (1971). Effect of tonsillectomy and adenoidectomy on nasopharyngeal antibody response to poliovirus. *New England Journal of Medicine*, **284**, 59–64.
Parsons, C. D., Thomas, M. H., and Garett, R. A. (1977). Colonic adenocarcinoma: a delayed complication of ureterosigmoidostomy. *Journal of Urology*, **118**, 31–4.
Penn, I. (1982). Surgically induced malignancies. In *Cancers induced by therapy*, (ed. I. Penn). *Cancer Surveys*, **1**, 745–61.
Petitti, D. B., Klein, R., Kipp, H., and Friedman, G. D. (1983). Vasectomy and the incidence of hospitalized illness. *Journal of Urology*, **129**, 760–2.
Pomare, E. W. and Heaton, K. W. (1973). The effect of cholecystectomy on bile salt metabolism. *Gut*, **14**, 753–62.
Raven, R. W. (1984). The surgeon and oncology. *Clinical Oncology*, **10**, 311–18.
Reddy, B. S. and Wynder, E. L. (1977). Metabolic epidemiology of colon cancer: fecal bile acids and neutral sterols in colon cancer patients and patients with adenomatous polyps. *Cancer*, **39**, 2533–9.
Roberts, J. P. and Taylor, I. (1990). Carcinoma of the breast associated with polyethylene strip augmentation. *Postgraduate Medical Journal*, **66**, 373–4.
Robinette, C. D. and Fraumeni, J. F., Jr (1977). Splenectomy and subsequent mortality in veterans of the 1939–45 war. *Lancet*, **ii**, 127–9.
Rosenberg, L., Palmer, J. R., and Shapiro, S. (1989). Vasectomy and prostate cancer. *American Journal of Epidemiology*, **130**, 829.
Rosenberg, S. A. (1989). Principles of surgical oncology. In *Cancer: principles and practice of oncology*, (ed. V. T. DeVita, Jr, S. Hellman, and S. A. Rosenberg), 3rd edn, pp. 236–46. Lippincott, Philadelphia.
Sandler, R. S., Johnson, M. D., and Holland, K. L. (1984). Risk of stomach cancer after gastric surgery for benign conditions: a case-control study. *Digestive Diseases and Sciences*, **29**, 703–8.
Schmähl, D., Thomas, C., and Auer, R. (1977). *Iatrogenic carcinogenesis*. Springer, Berlin.
Sheldon, C. A., McKinley, C. R., Hartig, P. R., and Gonzalez, R. (1983). Carcinoma at the site of ureterosigmoidostomy. *Diseases of the Colon and Rectum*, **26**, 55–8.

Sidney, S. (1987). Vasectomy and the risk of prostatic cancer and benign prostatic hypertrophy. *Journal of Urology*, **138**, 795–7.

Silverstein, M. J., *et al.* (1988). Breast cancer in women after augmentation mammoplasty. *Archives of Surgery*, **123**, 681–5.

Sooriyaarachchi, G. S., Johnson, R. O., and Carbone, P. P. (1977). Neoplasms of the large bowel following ureterosigmoidostomy. *Archives of Surgery*, **112**, 1174–7.

Stalsberg, H. and Taksdal, S. (1971). Stomach cancer following gastric surgery for benign conditions. *Lancet*, **ii**, 1175–7.

Stevenson, S., Hohn, R. B., Pohler, O. E. M., Fetter, A. W., Olmstead, M. L., and Wind, A. P. (1982). Fracture-associated sarcoma in the dog. *Journal of the American Veterinarian Medical Association*, **180**, 1189–96.

Strader, C. H., Weiss, N. S., and Daling, J. R. (1988). Vasectomy and the incidence of testicular cancer. *American Journal of Epidemiology*, **128**, 56–63.

Sunderman, F. W. (1989). Carcinogenicity of metal alloys in orthopedic prostheses: clinical and experimental studies. *Fundamental and Applied Toxicology*, **13**, 205–16.

Turunen, M. J. and Kivilaakso, E. O. (1981). Increased risk of colorectal cancer after cholecystectomy. *Annals of Surgery*, **194**, 639–41.

Urdaneta, L. F., Duffell, D., Creevy, C. D., Bradley Aust, J. (1966). Late development of primary carcinoma following ureterosigmoidostomy: report of three cases and literature review. *Annals of Surgery*, **164**, 503–13.

van Leeuwen, F. E., Somers, R., and Hart, A. A. M. (1987). Splenectomy in Hodgkin's disease and second leukaemias. *Lancet*, **ii**, 210–11.

Vennin, P., Bonneterre, J., and Demaille, A. (1985). Les cancers induits par la chirurgie existent-ils? *Le Concours Médical*, **107**, 1027–35.

Vernick, L. J. and Kuller, L. H. (1981). Cholecystectomy and right-sided colon cancer: an epidemiological study. *Lancet*, **ii**, 381–3.

Vernick, L. J., Kuller, L. H., Lohsoonthorn, P., Rycheck, R. R., and Redmond, C. K. (1980). Relationship between cholecystectomy and ascending colon cancer. *Cancer*, **45**, 392–5.

Vessey, M. P., Doll, R., Norman-Smith, B., and Hill, I. D. (1979). Thymectomy and cancer: a further report. *British Journal of Cancer*, **39**, 193–5.

Vianna, N. J., Greenwald, P., and Davies, J. N. P. (1971a). Tonsillectomy and Hodgkin's disease: the lymphoid tissue barrier. *Lancet*, **i**, 431–2.

Vianna, N. J., Greenwald, P., and Davies, J. N. P. (1971b). Nature of Hodgkin's disease agent. *Lancet*, **i**, 733–6.

Viste, A., *et al.* (1986). Risk of carcinoma following gastric operations for benign disease: a historical cohort study of 3470 patients. *Lancet*, **ii**, 502–5.

Vobecky, J., Caro, J., and Devroede G. (1983). A case-control study of risk factors for large bowel carcinoma. *Cancer*, **51**, 1958–63.

Werner, B., de Heer, K., and Mitschke, H. (1977). Cholecystectomy and carcinoma of the colon: an experimental study. *Zeitung für Krebsforschung*, **88**, 223–30.

Wolf, B. C. and Neiman, R. S. (1989). *Disorders of the spleen*, pp. 20–29. Saunders, Philadelphia.

8
Reducing the risk

PATRICIA FRASER

Introduction

As the treatment of certain cancers becomes more effective, and survival is prolonged, the risk of inducing a second, treatment-related malignancy has become an increasingly important consideration in planning the management of patients with cancer. Information on the relative carcinogenicity of different therapeutic agents will thus be of considerable value to clinicians in choosing or modifying treatment regimes so as to minimize this risk. In this chapter, relevant information principally from studies of new malignancies in long-term cancer survivors will be presented. Methods of surveillance and analysis which enable risks to be recognized, and ways in which treatment regimes may be modified to reduce them, will be discussed.

It has been known since the late 1950s that radiotherapy can both cure and cause cancer. During the last decade, it has become established that many chemotherapeutic agents, widely used in clinical medicine both to treat cancer and as immunosuppressive and anti-inflammatory agents, can also induce malignancies (Penn 1982). Both radiotherapy and chemotherapy had been in use for many years before it was realized that leukaemia and, later, other malignancies could arise as a consequence of treatment. Cancer may also occur as a late complication of some surgical procedures.

Although the assessment of carcinogenic risk to humans must rely to a large extent on data obtained from animal experiments, the interpretation of such data must always include an element of extrapolation from animals to man. Directly relevant information can be obtained, however, from systematic monitoring of human patients who have been treated for cancer, or other diseases, with radiotherapy and/or intensive chemotherapy. Such patients are among the few human beings to have been deliberately exposed to known carcinogens in large and carefully measured doses. If long-term follow-up can be assured, they provide an invaluable opportunity for quantitative assessment of carcinogenic risk.

Cancer treatment regimes are complex and are frequently revised, sometimes with the addition of new drugs. While a leukaemogenic effect may become apparent within 4 to 6 years, any treatment-related increase in the incidence of solid tumours is unlikely to be manifest for a decade or more, during which time the treatment regimes may have changed considerably.

For these risks to be clearly identifiable, accurate and complete clinical records are essential. Cancer patients should be under continuous surveillance, not just for the management of their illness, but also for possible late effects of their therapy for it. Careful documentation in hospital records of all treatment and of any subsequent malignancies will provide the basic data needed to establish which agents are carcinogenic (either alone or in combination), which patients may be most at risk, and what the magnitude of that risk may be. Hospital records are also the major source documents for population statistics on cancer incidence. For this reason, the development of efficient arrangements for prompt and complete notification of all malignancies to population-based cancer registries should be strongly encouraged. Cancer registry data can provide unbiased estimates of the risk of new malignancy, and can alert researchers to the need for more detailed study of increases in specific cancers which may be treatment-related.

In recent years, epidemiological methods have been used increasingly to study the carcinogenicity of therapeutic irradiation (Boice and Fraumeni 1984) and of drugs used in the treatment of patients with cancer and certain non-malignant conditions (Kaldor *et al.* 1986). Such studies are valuable in identifying for clinicians those therapeutic regimes which have proved either particularly safe and efficacious, or particularly hazardous in terms of long-term risk to patients. Three sources of data have been exploited in epidemiological studies of second malignancies in relation to therapy: population-based cancer registries, hospital-based cancer registries or case series, and clinical trials. All three sources provide an opportunity for surveillance of patients with a specified first cancer with a view to identifying the occurrence of a second. All three have their strengths and weaknesses.

Population surveillance

Measurement of the risk of a second primary malignancy in a cohort of patients with a specific first cancer depends on a comparison between the number of new malignancies observed in the cohort and the number which might have been expected on the basis of appropriate cancer incidence rates in the general population. The magnitude of this risk is difficult to assess from patients in hospital series, for even large hospitals will see second primary malignancies infrequently, and these are unlikely to be a representative sample of all second cancers. The population from which a hospital series is drawn may also be difficult to define. Population-based cancer registries, however, with their potential for recording all incident cancers in residents of a clearly defined catchment area, with efficient record linkage techniques and long-term follow-up, can provide unbiased estimates of the risk of leukaemia and solid tumours in patients with a given first cancer. This type of study has the additional benefit of identifying for more detailed investigation—e.g. in case-control studies—those patients with second cancers, some of which might be treatment-related.

In theory, population-based cancer registries offer the possibility of identifying all new cancers in their territory. In practice, the completeness of registration is variable, but the yield of second primary malignancies is still likely to be much greater, and more representative of the true distribution of second malignancies, than the yield obtained from hospital case series or clinical trials being carried out in the same area. Even so, unless the population it covers is very large, the number of second malignancies recorded by a single registry will usually be small, resulting in risk estimates with wide confidence limits, and too few cases for useful further study. Collaborative studies between several cancer registries, however, can lead to more precise overall estimates of second cancer risk in cohort studies, and can identify sufficient numbers of cases for the role of radiotherapy and cytostatic agents to be evaluated in subsequent case-control studies. Population-based cancer registries often record the initial treatment modality, but seldom obtain details of drugs, dose or subsequent treatment. Thus researchers using registry material as a starting point must collaborate closely with hospital staff to obtain the necessary therapeutic detail.

Several examples of collaborative studies between population-based cancer registries, and between registries and clinics, have now been published. The International Agency for Research on Cancer (IARC) has been largely responsible for coordinating the international collaborative effort. The approach has been to carry out cohort studies of individuals diagnosed with certain index cancers, in order to ascertain the incidence of second cancers, and then to evaluate in case-control studies the role of specific aspects of therapy, especially radiotherapy and cytostatic agents, by comparing the treatment records of individuals who developed a second primary malignancy with those of patients who did not.

Day and Boice (1983) and Boice *et al.* (1985) used a large collaborative group of mainly population-based cancer registries to study second cancers following cervical cancer, with particular reference to the effects of radiotherapy. By combining this registry material with other hospital series, the resulting study population was large enough to quantify the risk of radiation-induced leukaemia and to provide further information on the nature of the relationship between dose and response (Boice *et al.* 1987). Risk of leukaemia increased with increasing radiation dose until average doses of about 400 rad (4 Gy) were reached, and then decreased at higher doses. A similar evaluation of dose-response relationships for solid tumours is the subject of a further report (Boice *et al.* 1988). Kaldor *et al.* (1987) reported the results of an international collaborative study among cancer registries on second malignancies following testicular cancer, ovarian cancer, and Hodgkin's disease. This study was carried out as a first step in identifying for more detailed investigation those second primary malignancies which might be related to the treatment of the three index cancers. The case-control studies which followed were large enough to permit detailed analyses of the risk and time pattern of leukaemia according to amount and type of chemotherapy received (Haas *et al.* 1987; Kaldor *et al.* 1990*a,b*). Curtis *et al.* (1984) also examined the risk of leukaemia associated with the first course of cancer treatment by analysing data from the registries that participate in the US National Cancer Institute's Surveillance, Epidemiology and End Results (SEER) programme.

Although these collaborative studies are the largest and most detailed investigations ever made of cancer induced by radiotherapy and cytostatic drugs, they do have limitations. First, they are based on hospital records which have been compiled over several decades, and as such, may concern treatment regimes which are no longer used. Second, cancer registries inevitably work with a delay of some months or even years after diagnosis so that ascertainment of second cancers is not up to date. Third, because the studies are historical and patients are not available for further tests, it is not possible to combine them with newer clinical or laboratory investigations which might be of value in predicting long-term risks.

Clinical surveillance

Hospital-based registries and hospital case series

Many large cancer treatment centres maintain registers of patients in which are recorded not only details of treatment but long-term follow-up information on relapse or recurrence of disease, the occurrence of subsequent malignancies, and survival. The quality and availability of therapy records is usually high, and for many diseases a wide range of treatments will have been prescribed over time, enabling evaluation of their relative carcinogenicity. For example, a study of childhood cancer based on the records of 13 medical centres participating in the Late Effects Study Group concluded that both radiotherapy and chemotherapy with alkylating agents for childhood cancer increase the subsequent risk of bone cancer (Tucker *et al.* 1987). More recently Tucker *et al.* (1988) have reported on the risk of second cancers after treatment for Hodgkin's disease in patients identified through a hospital-based tumour registry; their data suggested that the risk of solid tumours after therapy for Hodgkin's disease continues to increase with time. Henry-Amar (1983) also reported on second cancers after radiotherapy and chemotherapy for early stage Hodgkin's disease in patients treated at institutions participating in the European Organisation for Research and Treatment of Cancer. These studies show that the value of information about therapy-related cancer can be considerably enhanced by

Clinical trials

The randomized prospective clinical trial has long been a model for the collection of valuable data on the use of therapeutic agents, usually to obtain unbiased assessments of their efficacy or short-term side effects. Randomized controlled trials also permit comparisons of cancer risk associated with different treatment regimes, however, as well as being free of the selection bias that may arise in observational studies. The treatments prescribed are highly standardized, clinical surveillance is intense and the quality of documentation is high. Provided compliance with treatment is good and long-term follow-up is assured, precise determination of the relative risk of any leukaemogenic or carcinogenic effects of the treatment can be obtained from these studies, in addition to answering the main question of which treatment is most effective. Such studies provide a unique opportunity to compare patients on a specific treatment regime with appropriate controls receiving different therapy for the same or similar neoplasms. The first formal comparison of the leukaemogenic potential of the widely used alkylating agents, melphalan and cyclophosphamide, was made in Britain in the Medical Research Council's first myelomatosis trial (Buckman et al. 1982). After a minimum follow-up of 12 years, six of the 258 trial participants had died of acute myeloid leukaemia, all after continuous treatment with melphalan and more than 4 years in the trial. This represented an incidence of acute myeloid leukaemia about 800 times that expected in 4-year survivors and demonstrated that melphalan is more leukaemogenic than cyclophosphamide in the treatment of patients with multiple myeloma.

The Environmental Epidemiology Branch of the National Cancer Institute (NCI) is systematically monitoring alkylating agents used in clinical trials for late complications, including cancer, in collaboration with the Division of Cancer Treatment, through which access is obtained to data from many NCI-supported randomized clinical trials (Greene 1984). As with observational studies, this ability to pool data from many centres has already proved beneficial. The occurrence of acute non-lymphocytic leukaemia among 1399 ovarian cancer patients treated in five randomized clinical trials was found to be confined to women exposed to alkylating agents, melphalan and chlorambucil individually contributing to the elevated risk (Greene et al. 1982). This study provided some evidence for a dose-response relationship. In nine randomized clinical trials of adjuvant therapy for gastro-intestinal cancer involving 3633 patients, a leukaemogenic effect of semustine (methyl-CCNU) was demonstrated (Boice et al. 1983, 1986). Of the 15 patients classified as having a leukaemic disorder, 14 had received semustine (RR = 12.4, 95 per cent CI 1.7–250). This study provides quantitative evidence that the nitrosoureas are leukaemogenic in humans and confirms previous observations that adjuvant therapy with alkylating agents may increase the risk of leukaemia. These large multi-centre trials provide an opportunity not only to evaluate dose-response relationships but also to examine the data for other risk factors, for example, age at treatment or concomitant radiotherapy or immunosuppression.

While case ascertainment in a registry-based study depends on a registration of leukaemia (or occasionally preleukaemia), hospital-based studies in which patient contact is maintained can use earlier endpoints. For example, in Boice et al.'s study (1983, 1986), suspected occurrences of leukaemia or preleukaemia were evaluated by inspection of bone marrow aspirates. Patients were classified as having acute non-lymphocytic leukaemia, acute myelodysplastic syndrome (preleukaemia 'in transition' to early overt leukaemia), simple preleukaemia, or a non-diagnostic marrow. Preleukaemia or acute myelodysplastic syndrome can be expected to develop into overt acute non-lymphocytic leukaemia within 1 or 2 years. Early recognition of the precursors of overt disease may be sufficiently timely for therapy to be modified or discontinued with subsequent recovery of the bone marrow. Detection of solid tumours at a premalignant or early stage will also generally offer the best prospect of cure.

Biomonitoring

In recent years there has been considerable progress in developing cytogenetic and molecular biological markers of dose or response to carcinogens (Bartsch et al. 1988). All the currently available markers reflect genetic damage at the level of DNA or the chromosome. They include carcinogen-DNA adducts, carcinogen-protein adducts (as a surrogate for DNA adducts), somatic cell mutations, micronuclei in lymphocytes, chromosomal aberrations, sister chromatid exchanges (SCE), unscheduled DNA synthesis, and oncogenic activation. Because the ability to cause genetic damage indicates potential carcinogenicity, such markers observed in humans may have significance regarding future cancer risk. However, while it is presumed that risk increases with dose of carcinogen, use of these markers to predict the actual magnitude of risk requires further evaluation. They should, therefore, be regarded as indicators of carcinogen exposure—not, in the present state of the art, as quantitative measures of risk.

Alkylating agents are known to induce chromosomal aberrations and SCE in the bone marrow cells and peripheral lymphocytes of cancer patients receiving chemotherapy (IARC 1981). SCE were counted in peripheral-blood lymphocytes from 10 patients with Behcet's syndrome receiving chlorambucil (Palmer et al. 1984): all 10 patients had abnormal

SCE counts. Damage was related both to daily dose and to duration of therapy, and was occurring at a greater rate than its repair. The patient who had been treated the longest and had received the highest dose had the highest SCE count, and died from acute leukaemia. SCE analysis may thus provide a method for detecting patients who are at particular risk from the oncogenic effects of cytostatic drugs.

Cytogenetic studies carried out on peripheral lymphocytes from cancer patients at different times after treatment with melphalan have indicated that measurement of SCE is useful for the study of newly-induced chromosome damage, but is less suitable for the detection of persistent, cytogenetic alterations long after therapy. On the other hand, the frequency of chromosomal aberrations may be increased for up to 10 years after melphalan therapy, the predominant aberrations being chromosomal translocations, marker chromosomes and cells with multiple complex arrangements (Lambert et al. 1986). These aberrations are compatible with cell survival and proliferation and are very rare in untreated subjects. The persistence of stable chromosomal rearrangements in peripheral lymphocytes after melphalan therapy may be an indication of an increased risk of second malignancy among these patients. In the rare chromosome instability syndromes (e.g. Fanconi's anaemia, ataxia telangectasia) an increased frequency of chromosomal aberrations is associated with a high incidence of cancer.

The SCE-inducing, chromosome-damaging, and carcinogenic effects of alkylating agents are probably associated with the DNA damage they induce. The events leading from DNA damage to chromosomal alterations, and possibly to cancer development are, however, far from clear. The non-random distribution of aberrations in the chromosomes of human tumours, and the finding that the chromosome breakpoints cluster at specific sites which often coincide with the location of cellular oncogenes (Yunis 1983; Mitelman 1984), suggest that chromosomal aberrations are important events in tumour development.

Methods are now available for quantifying carcinogen-DNA adducts in human tissues utilizing highly specific antibodies in sensitive immunoassays. The most commonly used methods involve competitive enzyme-linked immunosorbent assays (ELISA), the enzyme substrate giving a coloured, fluorescent or radiolabelled product. The development of methods for measuring the levels of specific carcinogen-DNA adducts in human samples, using monoclonal antibodies and highly sensitive fluorescence and post-labelling techniques, has provided a promising new tool for risk assessment and cancer prevention (Perera 1987; Bartsch et al. 1988).

The raising of polyclonal and/or monoclonal antibodies to DNA modified by chemotherapeutic agents provides a means of monitoring their interaction with critical cellular constituents. For example, a panel of monoclonal antibodies has been generated which specifically recognize DNA that has been modified by methoxsalen and ultraviolet (UV-A) light

(Santella *et al.* 1988). Methoxsalen plus UV-A light (PUVA) has been used clinically in the treatment of psoriasis, cutaneous T-cell lymphoma, and autoimmune disorders. *In vitro* studies of cultured human cells have shown that the photoadducts formed with DNA when methoxsalen is photo-activated with UV-A light are mutagenic and inhibit DNA synthesis. In patients treated with PUVA, a dose-dependent risk of cutaneous squamous cell carcinoma suggests that these photoadducts may also be carcinogenic (Stern *et al.* 1984). The quantification of methoxsalen adducts in patients with these techniques may thus provide a basis for estimating the cancer risk from treatment with PUVA.

Biological evidence that the mutagenicity and/or carcinogenicity of cis-platin is related to its genetic activity has led to measurement of cisplatin-DNA adducts in clinical studies (Poirier *et al.* 1988). Antibodies to these adducts have been produced and are being applied in serial sampling of lymphocyte DNA from cancer patients. Such methods may prove useful in monitoring response to anticancer therapy and in individualizing treatment regimes. While the current application is in clinical biomonitoring and in the validation of DNA adducts as measures of exposure and possible indicators of treatment efficacy, prospective epidemiological studies could be carried out to relate cisplatin-DNA adducts to future risk of a second cancer.

At present, a finding of high levels of adducts or cytogenetic markers in an exposed group of individuals indicates that they have received a significant genetic dose of a chemical and are potentially at increased risk of cancer. Such a group is a candidate for increased surveillance and even preventive measures to limit or to eliminate exposure. Unfortunately, interpretation of dose markers as indicators of cancer risk must currently be confined to the group rather than the individual, because of the lack of data definitely associating available markers with cancer risk and the observed overlap in levels of markers such as DNA adducts and SCE between exposed and unexposed groups. Only when correlations between markers and cancer risk have been established, and specific concentrations of markers that connote high risk have been identified, will it be possible to move to the individual level for risk prediction. Once they have been validated, however, these markers of preclinical response to carcinogenic exposures, reflecting very early events in the pathogenic process, could replace tumour incidence as endpoints. Such a development would be important for two reasons. First, it would enable much more rapid assessment of the carcinogenicity of a new treatment than is currently possible, because it would not be necessary to 'wait' for new cancers to develop. Second, provided that the biomarker used reflects a sufficiently early stage in cancer development, there would also be the possibility of intervention to prevent the tumours from actually occurring.

In this context, it has recently been proposed that measurements of DNA damage should be included in large-scale, prospective studies of long-term cancer survivors (Kaldor and Day 1988). These studies will involve collaboration between the epidemiological units at IARC and the European Organization for Research and Treatment of Cancer which can offer clinical expertise and access to patients in many hospitals and clinics. If some early marker of second cancer risk can be found, which can be used to identify patients who appear to be placed at higher risk by their treatment, appropriate modifications aimed at reducing risk could be applied.

Modification of treatment regimes

Some forms of treatment have already been abandoned following recognition of their carcinogenicity (IARC 1980). For example, it is now generally accepted that arsenic by mouth caused cancer of the skin, that analgesic mixtures containing phenacetin can cause tumours of the renal pelvis, and that intrauterine exposure to diethylstilboestrol predisposes to the development of clear cell carcinoma of the vagina (Birch, Chapter 6, this volume). Prescribing practices have also been influenced by the realization that use of unopposed conjugated oestrogens for the relief of postmenopausal symptoms is associated with an increased risk of endometrial cancer.

Surgical procedures

Malignancies may also occur as late complications of several surgical procedures (Penn 1982; Coleman, Chapter 7, this volume). For example, stomach cancer may develop after gastric surgery for benign disease, and carcinoma of the colon may arise as a late complication of operations resulting in prolonged exposure of the intestine to urine. The risk of such surgically-induced malignancies may be reduced by operative procedures which avoid reflux of duodenal contents into the stomach, or urinary diversion into the colon. The present trend of performing less radical surgery for breast cancer should reduce still further the already very low incidence of post-mastectomy lymphangiosarcoma, a rare complication of lymphoedema.

It has been reported recently that splenectomy in patients with Hodgkin's disease greatly increases the risk of leukaemia (van Leeuwen *et al.* 1987; Kaldor *et al.* 1990*b*). Though a staging laparotomy with removal of the spleen is an accepted procedure in the management of Hodgkin's disease, its popularity in recent years has declined. A reduction in the risk of leukaemia associated with splenectomy can therefore be expected.

Immunosuppressive therapy

In transplant surgery, immune responses must be altered to obtain acceptance of the grafts, if histocompatibility differences exist between host and donor. It is well recognized that drug-induced suppression of immunity in the graft recipient (host), usually with azathioprine and corticosteroids or cyclosporin, is associated with an increased incidence of certain malignancies, in particular lymphomas which have a short latent period and a predilection for the central nervous system (Kinlen 1982). The same cancers that occur in excess in transplant recipients also show an excess in patients without transplants who have been treated with immunosuppressive drugs for other reasons. This implies that the excesses in transplant recipients are not simply due to immunostimulation of the recipient by foreign antigens arising from the graft, but that the immunosuppressive therapy itself is an important factor (Kinlen 1985). Some workers have reported regression of post-transplant lymphoproliferative neoplasms, without rejection of the graft, following reduction or discontinuation of immunosuppressive therapy (Starzl et al. 1984). There is clearly a need for immune control methods that will induce specific non-reactivity in the host to the foreign histocompatibility antigens of the allograft, but will otherwise leave the host's immune defences intact. Some recent advances along these lines have included the use of antilymphocytic serum to combat rejection, and an increasing potential for monoclonal anti-T cell antibodies (Morris and Tilney 1986).

The use of immunosuppressive drugs in the treatment of rheumatoid arthritis and other immunoinflammatory disorders should probably be reserved for patients with severe progressive disease uncontrolled by conventional therapy, or else for life-threatening complications. Many of the malignancies associated with immunosuppression are also tumours for which evidence of a viral aetiology is strongest. It has been suggested therefore that anti-viral agents, such as acyclovir, may have a role in reducing the risk of lymphoproliferative malignancies in immunosuppressed patients (Hanto et al. 1981).

Radiotherapy

Radiation therapy is no longer the treatment of choice for ankylosing spondylitis, tinea capitis or benign gynaecological disorders, because of the risk of inducing a malignancy. Leukaemia and various solid tumours have also been reported after radiotherapy for cancers of the breast, ovary, endometrium, cervix and thyroid, Hodgkin's disease, and non-Hodgkin's lymphoma (Boice 1988). Therapeutic irradiation remains one of the mainstays of treatment for these malignancies, however, and improved techniques of radiotherapy are being used to minimize the risk to non-target organs. The absolute excess risk to such organs is, in any case, not very large, even though for some combinations of first and second malignancy, the relative risk is quite high (Day, Chapter 2, this volume). The introduction of megavoltage radiation, improved tumour imaging, and localization, and more efficient focusing and dose fractionation, for example, have enabled the side effects of radiotherapy to be reduced, while still achieving greater tumour resolution. It may be assumed that subsequent cancer risk after X-irradiation will also be reduced by these means, but there is unlikely to be good evidence of such an improvement for some years, and the increasing use of adjuvant chemotherapy may make it difficult, if not impossible, to identify clearly any such reduction in risk.

Cytostatic therapy

Clinical approaches to reducing the carcinogenicity of regimes which involve cytostatic agents have focused on modification of either the 'cocktail' or combination of drugs used, or the dose schedule, or the mode of administration. In the laboratory, attempts have also been made to develop safer and more effective anti-cancer drugs. Some examples of these approaches to reducing the risk of new malignancies after cytostatic therapy are described below. For a comprehensive review of the carcinogenicity of cytostatic drugs, including the modification of treatment regimes, the reader is referred to Schmähl and Kaldor (1986).

Avoidance of cytostatic drugs It is becoming apparent that the use of cytostatic agents for the treatment of non-neoplastic disease, if unavoidable, should be confined to patients in whom all other treatment has failed, or whose condition is life-threatening. Some therapy-induced leukaemias have arisen when relatively low doses of cytostatic agents have been used as adjuvant therapy to surgical treatment of cancers of the breast, lung, and gastrointestinal tract (Lerner 1977; Stott *et al.* 1977; Boice *et al.* 1983). Even in cancer patients, therefore, adjuvant cytostatic therapy should be used cautiously, preferably in carefully controlled trials. Indeed Schmähl and Kaldor (1986) have advised that

> in adjuvant chemoprevention, no carcinogenic drug should be administered. This applies in particular to those cases in which a high percentage of cures can be achieved by means of classical methods (operation, irradiation) ... Furthermore, particular restraint should be exercised in young patients, in whom possible late effects of treatment with carcinogenic drugs are more likely to manifest themselves.

Since there is mounting evidence that risk increases with total cumulative dose (Pedersen-Bjergaard *et al.* 1987; Kaldor *et al.* 1990*a,b*), the need for long-term maintenance chemotherapy in the treatment of malignancy should be very carefully assessed.

Cytostatic agents are indispensable in the treatment of many malignancies because they save or prolong life. Fortunately, despite their widespread use in cancer therapy, the development of a second primary malignancy still remains a relatively uncommon late complication, arising in perhaps 2–11 per cent of patients (Penn 1986). Against this background, the use of a number of cytostatic agents with known carcinogenic potential is justified, but it now requires the clinician to make a more sober calculation of risk (of new cancers) and benefit (remission or cure) than was thought necessary even 10 or 15 years ago. Two alkylating agents have proved to be too carcinogenic altogether, namely chlornaphazine (Thiede *et al.* 1964) and treosulphan (Pedersen-Bjergaard *et al.* 1980), and they are no longer used in clinical chemotherapy at all.

Modification of components of treatment regimes Since many cytostatic agents have therapeutic benefits, it is clearly desirable to minimize their long-term danger to patients. Efforts in this direction have become manifest in increasingly strict prescribing practices. One strategy for the prevention of further malignancies is to use single drugs or combinations of drugs that have no (or a comparatively low) carcinogenic potential while maintaining therapeutic benefit. Calabresi (1983) drew attention to several drug combinations for which beneficial results have been demonstrated without any apparent increased risk of a second malignancy. The combination of cisplatin, vinblastine, and bleomycin as curative treatment for disseminated testicular cancer was among those cited. Patients with Hodgkin's disease treated with ABVD (adriamycin, bleomycin, vinblastine, and dacarbazine) also appear to be at lower risk than patients treated with the standard MOPP regime (nitrogen mustard, vincristine (Oncovin), procarbazine, and prednisone) (Valagussa *et al.* 1982). However, there is still an absence of extensive follow-up data on ABVD patients, and a leukaemogenic effect cannot yet be excluded (Kaldor *et al.* 1990*b*). In patients with ovarian cancer, the combination of adriamycin and cisplatin was associated with significantly increased leukaemia risk (Kaldor *et al.* 1990*a*).

Alkylating agents and other drugs that bind tightly to DNA have, most frequently, been associated with the appearance of new malignancies (IARC 1981). Berger *et al.* (1983) substituted vincristine, an antimitotic drug, for cyclophosphamide, an alkylating agent with known carcinogenic activity, in combination with methotrexate and 5-fluorouracil. This combination was free of carcinogenic potential in experimentally-induced mammary carcinoma in rats, and its therapeutic effect was equivalent to that of the combination containing cyclophosphamide. These results suggest that drug combinations which do not include alkylating agents may prove to be less carcinogenic than those which do, yet without compromising the results currently being obtained in the treatment of various haematological and lymphoid malignancies. Because of high risks of acute leukaemia, Pedersen-Bjergaard *et al.* (1987) have advised against using two or more alkylating agents in combination, especially in older patients, and against retreatment with alkylating agents for patients who relapse with Hodgkin's disease.

The carcinogenicity in animals of alkylating cytostatic drugs is well known, and for many such compounds demonstration of carcinogenicity in animals preceded the observation in humans (Berger 1986). However, extrapolation from animal experimental results to clinical practice is not straightforward. To evaluate the predictive value of animal tests in assessing the risk of second cancer in patients treated with antineoplastic drugs, Kaldor *et al.* (1988) examined the correlation between the carcinogenic potency of antineoplastic drugs in humans and rodents. For the limited number of agents for which data were available, a reasonable correlation in potency ranking was observed, cyclophosphamide, chlorambucil, and melphalan having very similar rank in humans and rodents. More extensive data warrant examination before it can be concluded that this method can be used reliably to predict which drugs present a greater carcinogenic risk to patients. This approach would, however, be particularly useful for assessing the carcinogenicity of agents which cannot be evaluated in humans, either because they are rarely given alone, or because they have been introduced too recently into clinical practice for appropriate studies to have been made.

Combined modality treatment with both radiotherapy and chemotherapy is a cornerstone of modern cancer therapy. Since both ionizing radiation and certain cytostatic agents are established carcinogens, the possibility exists that administering both concurrently or sequentially might pose a greater risk of a subsequent malignancy than using either modality alone. The picture is not clear, as varying effects have been described. Penn (1986) reviewed a number of reports which strongly suggest that the risk increases if radiotherapy is administered in addition to cytostatic therapy. For example, second cancers were substantially increased in patients with Hodgkin's disease, non-Hodgkin's lymphoma, and Ewing's sarcoma who received both treatments. In contrast, in a large international collaborative case-control study of leukaemia following Hodgkin's disease (Kaldor *et al.* 1990*b*) and ovarian cancer (Kaldor *et al.* 1990*a*), radiotherapy in combination with chemotherapy did not increase the risk of leukaemia over that produced by chemotherapy alone. In one study actinomycin-D was reported to have decreased the risk of radiation-induced cancers in children who received both therapies (D'Angio *et al.* 1976), but in another there was no evidence that the joint effect of actinomycin-D and radiation therapy was protective against the development of thyroid cancer (Boice 1988).

Alkylating agents are more potent carcinogens than therapeutic irradiation (Greene 1984). Until more is known about the interaction between radiotherapy and chemotherapy in human carcinogenesis, intensive chemotherapy should probably not be combined with intensive radiotherapy except when the extent and severity of disease pose an immediate threat to the patient's life. Accurate staging of Hodgkin's disease may avoid unnecessary combination therapy; radiotherapy alone is highly effective in the treatment of localized Hodgkin's disease and it is relatively free of troublesome side effects.

Urotoxic side effects have been a limiting factor for the therapeutic use of cyclophosphamide and other oxazaphosphorine cytostatics. The sterile haemorrhagic cystitis that frequently occurs in patients treated with cyclophosphamide is due to the release of acrolein from 4-hydroxycyclophosphamide accumulating in the bladder, and this can be prevented by concurrent administration of mesna (sodium 2-mercaptoethane sulphonate). In extensive experiments on rats, it has been demonstrated that the cyclophosphamide-induced occurrence of urinary bladder tumours can be reduced or even eliminated by simultaneous administration of mesna (Brock and Pohl 1986). The successful prevention of acute urotoxic effects, particularly of haemorrhagic cystitis, by mesna has raised the hope that such treatment might help to reduce or even to prevent the risk of bladder cancer resulting from long-term treatment with cyclophosphamide. This is the first example of the rational development of an antidote against the carcinogenic activity of an alkylating agent. The abolition of toxic damage to the bladder and kidney makes it possible to administer oxazaphosphorine cytostatics at higher doses, and thus to improve their potential therapeutic efficacy considerably. Preliminary experiments with cisplatin and mesna indicate that kidney tumours caused by cisplatin might also be prevented by additional medication with mesna (Petru and Schmähl 1987).

Modification of dose and mode of administration Many other contemporary anti-cancer drugs have a narrow therapeutic index, in that acute toxicity of the drug for normal tissues frequently limits the amount of drug which can be safely administered in a single dose, as well as the interval between treatments, and the cumulative dose which can be given. Clinicians have experimented with modification to dose schedules with the aim of maximizing their efficacy while minimizing both acute and late toxic effects.

Cisplatin is a good example: this drug is active against many human cancers, but side effects such as severe nausea and vomiting, peripheral neuropathy, ototoxicity, myelosuppression, and renal toxicity often limit the tolerable dose. To increase delivery of cisplatin to the tumour while maintaining tolerable exposure of normal tissues, intra-arterial administration has been used. Initial clinical studies demonstrated increased response rates in patients with melanoma, soft tissue sarcoma, colon carcinoma, and both primary and secondary malignant tumours of the brain. In an attempt to obtain even greater tumour exposure without increased systemic toxicity, Oldfield *et al.* (1987) combined intra-arterial infusion with haemodialysis of the jugular venous drainage during intracarotid chemotherapy in patients with malignant gliomas. The application of this imaginative technique and the related pharmacokinetic principles should clearly be studied further, both with other drugs and for cancers in other parts of the body where the regional venous drainage is readily accessible.

There are conflicting animal data as to whether cytostatic agents are carcinogenic because of their immunosuppressive properties or because of a direct effect on cells, or both. The type and severity of therapy-induced immunosuppression depend on the drug or drug combination used: alkylating agents and antimetabolites are intensely immunosuppresive, antibiotics, and antimitotics less so (Obrecht and Obrist 1986). Depression of immunity is more profound with continuous, prolonged drug administration than with intermittent therapy, which gives the lymphoid tissues a chance to recover between courses of treatment (Hersh *et al.* 1973). Initial experimental results point to the possibility that changes in the timing of treatment regimens could reduce the risk of developing a drug-induced second malignancy (Schmähl and Kaldor 1986). Obrecht and Obrist (1986) give some examples of successful reduction of intensity and duration of therapy in men with non-seminoma testicular cancer and children with acute lymphoblastic leukaemia.

Avoidance of other risk factors Cytostatic agents may act synergistically with various carcinogens to which the patient may be exposed, such as tobacco smoke, sunlight and other physical and chemical agents. Because of the susceptibility of patients with Hodgkin's disease to subsequent lung cancer, cessation of smoking should be viewed as an essential component of the treatment for Hodgkin's disease. Careful surveillance of dysplastic naevi is also needed, particularly since the melanomas that arise after Hodgkin's disease tend to be aggressive (Tucker *et al.* 1988).

Modification of drug structure and drug targeting Many drugs that kill tumour cells by interacting directly with their DNA can also cause non-lethal mutations which may result in a second primary malignancy. Alkylating agents have a non-selective mechanism of action, and can alkylate all molecules that contain nucleophilic sites (Connors 1986). Because tissues with a larger number of proliferating cells are particularly sensitive, the commonly used alkylating agents are especially toxic to bone marrow. Several techniques have been developed to circumvent or decrease the myelosuppressive effects of antineoplastic drugs (Testa and Gale, 1988). These include, for example, bone marrow autotransplantation or rescue with autologous cryopreserved bone marrow cells, and new systems of drug delivery using liposome-entrapped drugs.

The type of DNA modification induced by alkylating agents can be exploited in the design of safer and more effective drugs. For example, by decomposing in different ways, the haloethylnitrosoureas generate intermediates that can introduce either hydroxyethyl or haloethyl groups into DNA. While haloethyl derivatives lead to cross-linked structures which may be cytotoxic, one hydroxyethyl derivative may be responsible for at least some of the carcinogenicity of these compounds (Ludlum 1986). It may therefore be possible to link certain DNA modifications to the carcinogenicity of nitrosoureas and other modifications to their cytotoxic/therapeutic activity.

The nitrosoureas in clinical use (BCNU, CCNU, and methyl-CCNU) are highly active anti-cancer agents, but their toxicity to bone marrow is dose-limiting, and is probably related to their leukaemogenicity. By contrast, the bone marrow toxicity associated with dacarbazine and related cytostatic triazenes is relatively moderate. Although dacarbazine is an established laboratory carcinogen, no malignancy has been reported in humans treated with dacarbazine alone. It appears experimentally that the carcinogenic and tumour-inhibitory effects of individual triazenes may be inversely correlated. They might therefore find wider clinical application if derivatives with less carcinogenic potential are proved to have increased anticancer activity in comparison with dacarbazine (Berger 1986).

The development of immunochemical methods to measure cisplatin-DNA adducts has made it possible to study the induction and repair of cisplatin-DNA lesions in cells. Information about the relative cytotoxicity (or carcinogenicity) caused by different lesions in DNA is of fundamental importance to chemotherapy (and cancer research) and cisplatin appears to offer an excellent model for such research (Forsti and Hemminki 1986). The types of products formed in the cross-linking reactions can be varied by chemical modifications of the platinum analogues, a feature exploited in structure-activity exercises for developing new anti-cancer agents.

Many human tumours contain hormone-receptors and attempts have been made to develop cytostatic agents that are effective in treating hormone-responsive tumours. Structure-activity studies have shown that the binding of alkylating/cross-linking anti-neoplastic agents to receptor carrier molecules is a promising way of targeting the drug on receptor-containing tumours, while sparing normal tissues which are sensitive to the drug but which do not contain the receptors (Eisenbrand *et al.* 1986). Carrier molecules for nitrosoureas are under investigation, and oestrophilic ligands have been introduced into platinum complexes in order to make them active specifically against hormone-dependent tumours. Cytostatic agents attached to monoclonal antibodies can be selectively targeted on cancer cells which possess antigen markers on the cell surface (Brock 1986). Enhanced tumour regression has been reported in a preliminary trial of specific targeting therapy against malignant glioma, using lymphokine-activated killer cells treated with bispecific antibody (Nitta *et al.* 1990). Even greater selectivity might one day be achievable if, rather than hitting randomly at all sites in the genome, a chemotherapeutic agent could be aimed specifically at the sites that control malignant growth, such as those related to oncogenes (Forsti and Hemminki 1986).

Conclusion

Recognition of the risk of new cancers as a consequence of chemotherapy, radiotherapy or surgery has led to attempts to modify treatment regimes so as to reduce the risk without compromising the good results now attainable in many patients with both neoplastic and non-neoplastic disease. Alternatives to surgical procedures where malignancy is a recognized late complication, and improved techniques of radiotherapy minimizing exposure of non-target organs, have achieved this end.

The very existence of the problem of new malignancies in cancer patients can be seen as a tribute to the success of chemotherapy, along with radiotherapy, in improving the survival of patients with cancer. However, the 40 or so drugs currently employed by medical oncologists for cancer therapy have limited areas of usefulness and considerable associated toxicity. Despite all the advances, the selectivity and therapeutic range of anti-cancer drugs currently available are by no means satisfactory (Brock 1986). The aim now should be towards individualization of treatment with safer and more effective drugs. This should follow from a better understanding of the modes of action of various anti-neoplastic agents, improved ability to target therapy on tumour cells, and prevention of side effects by regional detoxification. In the light of recent developments, it now seems reasonable to expect that novel ways of monitoring the patient's response to treatment and the degree of damage to his or her DNA may ultimately lead to the early identification of those patients who are at higher risk as a result of their treatment, and in whom timely intervention may prevent new tumours from occurring.

References

Bartsch, H., Hemminki, K., and O'Neill, I. K. (ed.) (1988). *Methods for detecting DNA damaging agents in humans: applications in cancer epidemiology and prevention*, IARC Scientific Publications No. 89. International Agency for Research on Cancer, Lyon.

Berger, M. R. (1986). Carcinogenicity of alkylating cytostatic drugs in animals. In *Carcinogenicity of alkylating cytostatic drugs*, IARC Scientific Publications No. 78, (ed. D. Schmähl and J. M. Kaldor), pp. 161–76. International Agency for Research on Cancer, Lyon.

Berger, M., Habs, M., and Schmähl, D. (1983). Noncarcinogenic chemotherapy with a combination of vincristine, methotrexate and 5-fluorouracil (VMF) in rats. *International Journal of Cancer*, **32**, 231-6.

Boice, J. D. Jr (1988). Carcinogenesis—a synopsis of human experience with external exposure in medicine. *Health Physics*, **55**, 621-30.

Boice, J. D. Jr and Fraumeni, J. F. Jr (ed.) (1984). *Radiation carcinogenesis: epidemiology and biological significance*, Progress in Cancer Research and Therapy, Vol. 26. Raven Press, New York.

Boice, J. D. Jr, et al. (1983). Leukaemia and preleukaemia after adjuvant treatment of gastrointestinal cancer with semustine (methyl-CCNU). *New England Journal of Medicine*, **309**, 1079-84.

Boice, J. D. Jr, et al. (1985). Second cancers following radiation treatment for cervical cancer. An international collaboration among cancer registries. *Journal of the National Cancer Institute*, **74**, 955-75.

Boice, J. D. Jr, et al. (1986). Leukaemia after adjuvant chemotherapy with semustine (methyl-CCNU)—evidence of a dose-response effect. *New England Journal of Medicine*, **314**, 119-20.

Boice, J. D. Jr, et al. (1987). Radiation dose and leukaemia risk in patients treated for cancer of the cervix. *Journal of the National Cancer Institute*, **79**, 1295-311.

Boice, J. D. Jr, et al. (1988). Radiation dose and second cancer risk in patients treated for cancer of the cervix. *Radiation Research*, **116**, 3-55.

Brock, N. (1986). Ideas and reality in the development of cancer chemotherapeutic agents, with particular reference to oxazaphosphorine cytostatics. *Journal of Cancer Research and Clinical Oncology*, **111**, 1-12.

Brock, N. and Pohl, J. (1986). Prevention of urotoxic side effects by regional detoxification with increased selectivity of oxazaphosphorine cytostatics. In *Carcinogenicity of alkylating cytostatic drugs*, IARC Scientific Publications No. 78, (ed. D. Schmähl and J. M. Kaldor), pp. 269-79. International Agency for Research on Cancer, Lyon.

Buckman, R., Cuzick, J., and Galton, D. A. G. (1982). Long-term survival in myelomatosis. *British Journal of Haematology*, **52**, 589-99.

Calabresi, P. (1983). Leukaemia after cytotoxic chemotherapy—a pyrrhic victory? *New England Journal of Medicine*, **309**, 1118-19.

Connors, T. A (1986). Antitumour alkylating agents: cytotoxic action and organ toxicity. In *Carcinogenicity of alkylating cytostatic drugs*, IARC Scientific Publications No. 78, (ed. D. Schmähl and J. M. Kaldor), pp. 143-5. International Agency for Research on Cancer, Lyon.

Curtis, R. E., Hankey, B. F., Myers, H. H., and Young, J. L. Jr (1984). Risk of leukaemia associated with the first course of cancer treatment: an analysis of the Surveillance, Epidemiology and End Results program experience. *Journal of the National Cancer Institute*, **72**, 531-44.

Day, N. E. and Boice, J. D. Jr (ed.) (1983). *Second cancer in relation to radiation treatment for cervical cancer: from the International Radiation Study Group on Cervical Cancer*, IARC Scientific Publications No. 52. International Agency for Research on Cancer, Lyon.

D'Angio, G. J., et al. (1976). Decreased risk of radiation-associated second malignant neoplasms in actinomycin-D-treated patients. *Cancer*, **37**, 1177-85.

Eisenbrand, G., Muller, N., Schreiber, J., Stahl, W., and Sterzel, W. (1986). Drug design: nitrosoureas. In *Carcinogenicity of alkylating cytostatic drugs*, IARC Scientific Publications No. 78, (ed. D. Schmähl and J. M. Kaldor), pp. 281–94. International Agency for Research on Cancer, Lyon.

Försti, A. and Hemminki, K. (1986). Unique properties of cisplatin in reactions with nucleophiles. In *Carcinogenicity of alkylating cytostatic drugs*, IARC Scientific Publications No. 78, (ed. D. Schmähl and J. M. Kaldor), pp. 101–10. International Agency for Research on Cancer, Lyon.

Greene, M. H. (1984). Interaction between radiotherapy and chemotherapy in human leukaemogenesis. In *Radiation carcinogenesis: epidemiology and biological significance*, (ed. J. D. Boice, Jr and J. F. Fraumeni, Jr), pp. 199–210. Raven Press, New York.

Greene, M. H., Boice, J. D. Jr, Greer, B. E., Blessing, J. A., and Dembo, A. (1982). Acute nonlymphocytic leukaemia following alkylating agent therapy for ovarian cancer—in a study of five randomised clinical trials. *New England Journal of Medicine*, **307**, 1416–21.

Haas, J. F., et al. (1987). Risk of leukaemia in ovarian tumour and breast cancer patients following treatment by cyclophosphamide. *British Journal of Cancer*, **55**, 213–18.

Hanto, D. W., et al. (1981). Clinical spectrum of lymphoproliferative disorders in renal transplant recipients and evidence for the role of Epstein–Barr virus. *Cancer Research*, **41**, 4253–61.

Henry-Amar, M. (1983). Second cancers after radiotherapy and chemotherapy for early stages of Hodgkin's disease. *Journal of the National Cancer Institute*, **71**, 911–16.

Hersh, E. M., et al. (1973). Host defence, chemical immunosuppression and the transplant recipient—relative effects of intermittent versus continuous immunosuppressive therapy with reference to the objectives of treatment. *Transplantation Proceedings*, **5**, 1191–5.

IARC Working Group (1980). An evaluation of chemicals and industrial processes associated with cancer in humans based on human and animal data: IARC Monographs, Volumes 1 to 20. *Cancer Research*, **40**, 1–12.

IARC (1981). *Monographs on the evaluation of the carcinogenic risk of chemicals to humans*, Vol. 26, *Some antineoplastic and immunosuppressive agents*. International Agency for Research on Cancer, Lyon.

Kaldor, J. M. and Day, N. E. (1988). Epidemiological studies of the relationship between carcinogenicity and DNA damage. In *Methods for detecting DNA damaging agents in humans: applications in cancer epidemiology and prevention*, IARC Scientific Publications No. 89, (ed. H. Bartsch, K. Hemminki, and I. K. O'Neill), pp. 460–8. International Agency for Research on Cancer, Lyon.

Kaldor, J. M., Day, N. E., and Shiboski, S. (1986). Epidemiological studies of anticancer drug carcinogenicity. In *Carcinogenicity of alkylating cystostatic drugs*, IARC Scientific Publications No. 78, (ed. D. Schmähl and J. M. Kaldor), pp. 189–201. International Agency for Research on Cancer, Lyon.

Kaldor, J. M., et al. (1987). Second malignancies following testicular cancer, ovarian cancer and Hodgkin's disease: an international collaborative study among cancer registries. *International Journal of Cancer*, **39**, 571–85.

Kaldor, J. M., Day, N. E., and Hemminki, K. (1988). Quantifying the carcino-

genicity of antineoplastic drugs. *European Journal of Cancer and Clinical Oncology*, **24**, 703–11.
Kaldor, J. M., *et al.* (1990a). Leukemia following chemotherapy for ovarian cancer. *New England Journal of Medicine*, **322**, 1–6.
Kaldor, J. M., *et al.* (1990b). Leukemia following Hodgkin's disease. *New England Journal of Medicine*, **322**, 7–13.
Kinlen, L. J. (1982). Immunosuppressive therapy and cancer. *Cancer Surveys*, **1**, 565–83.
Kinlen, L. J. (1985). Incidence of cancer in rheumatoid arthritis and other disorders after immunosuppressive treatment. *American Journal of Medicine*, **78** (Suppl. 1A), 45–9.
Lambert, B., Holmberg, K., and Einhorn, N. (1986). Chromosome damage and second malignancy in patients treated with melphalan. In *Carcinogenicity of alkylating cytostatic drugs*, IARC Scientific Publications No. 78, (ed. D. Schmähl and J. M. Kaldor), pp. 147–60. International Agency for Research on Cancer, Lyon.
Lerner, H. (1977). Second malignancies diagnosed in breast cancer patients while receiving adjuvant chemotherapy at the Pennsylvania Hospital. *Proceedings of the American Association for Cancer Research*, **18**, 340.
Ludlum, D. B. (1986). Nature and biological significance of DNA modification by the haloethylnitrosoureas. In *Carcinogenicity of alkylating cytostatic drugs*, IARC Scientific Publications No. 78, (ed. D. Schmähl and J. M. Kaldor), pp. 71–81. International Agency for Research on Cancer, Lyon.
Mitelman, F. (1984). Restricted number of chromosomal regions implicated in aetiology of human cancer and leukaemia. *Nature*, **310**, 325–27.
Morris, P. J. and Tilney, N. L. (1986). *Progress in transplantation*, Vol. 3. Churchill Livingstone, Edinburgh.
Nitta, T., Sato, K., Yagita, H., Okumura, K., and Ishii, S. (1990). Preliminary trial of specific targeting therapy against malignant glioma. *Lancet*, **335**, 368–71.
Obrecht, J. P. and Obrist, R. (1986). Clinical use of chemotherapeutic agents. In *Carcinogenicity of alkylating cytostatic drugs*, IARC Scientific Publications No. 78, (ed. D. Schmähl and J. M. Kaldor), pp. 37–51. International Agency for Research on Cancer, Lyon.
Oldfield, E. H., *et al.* (1987). Reduced systemic drug exposure by combining intraarterial *cis*-diamminedichloroplatinum (II) with haemodialysis of regional venous drainage. *Cancer Research*, **47**, 1962–7.
Palmer, R. G., Dore, C. J., and Denman, A. M. (1984). Chlorambucil-induced chromosome damage to human lymphocytes is dose-dependent and cumulative. *Lancet*, **i**, 246–9.
Pedersen-Bjergaard, J., *et al.* (1980). Acute non-lymphocytic leukemia in patients with ovarian carcinoma following long-term treatment with treosulfan (= dihydroxybusulfan). *Cancer*, **45**, 19–29.
Pedersen-Bjergaard, J., *et al.* (1987). Risk of therapy-related leukaemia and pre-leukaemia after Hodgkin's disease. *Lancet*, **ii**, 83–8.
Penn, I. (1982). Cancers induced by therapy. *Cancer Surveys*, **1**, 565–782.
Penn, I. (1986). Malignancies induced by drug therapy: a review. In *Carcinogenicity of alkylating cytostatic drugs*, IARC Scientific Publications No. 78, (ed. D.

Schmähl and J. M. Kaldor), pp. 13–27. International Agency for Research on Cancer, Lyon.
Perera, F. P. (1987). Molecular cancer epidemiology: a new tool in cancer prevention. *Journal of the National Cancer Institute*, **78**, 887–98.
Petru, E. and Schmähl, D. (1987). Second malignancies—risk reduction. *Cancer Treatment Reviews*, **14**, 337–43.
Poirier, M. C., Egorin, M. J., Fichtinger-Schepman, A. M. J., Yuspa, S. H., and Reed, E. (1988). DNA adducts of cisplatin and carboplatin in tissues of cancer patients. In *Methods of detecting DNA damaging agents in humans: applications in cancer epidemiology and prevention*, IARC Scientific Publications No. 89, (ed. H. Bartsch, K. Hemminki, and I. K. O'Neill), pp. 313–20. International Agency for Research on Cancer, Lyon.
Santella, R. M., Yang, X. Y., DeLeo, V. A., and Gasparro, F. P. (1988). Detection and quantification of 8-methoxypsoralen-DNA adducts. In *Methods for detecting DNA damaging agents in humans: applications in cancer epidemiology and prevention*, IARC Scientific Publications No. 89, (ed. H. Bartsch, K. Hemminki, and I. K. O'Neill), pp. 333–40. International Agency for Research on Cancer, Lyon.
Schmähl, D. and Kaldor, J. M. (ed.) (1986). *Carcinogenicity of alkylating cytostatic drugs*, IARC Scientific Publications No. 78. International Agency for Research on Cancer, Lyon.
Starzl, T. E., *et al.* (1984). Reversibility of lymphomas and lymphoproliferative lesions developing under cyclosporin-steroid therapy. *Lancet*, **i**, 583–7.
Stern, R. S., Laird, N., Melski, J., Parrish, J. A., Fitzpatrick, T. B., and Bleich, H. L. (1984). Cutaneous squamous-cell carcinoma in patients treated with PUVA. *New England Journal of Medicine*, **310**, 1156–61.
Stott, H., Fox, W., Girling, D. J., Stephens, R. J., and Galton, D. A. G. (1977). Acute leukaemia after busulphan. *British Medical Journal*, **ii**, 1513–17.
Testa, N. G. and Gale, R. P. (ed.) (1988). *Hematopoiesis: long-term effects of chemotherapy and radiation*. Dekker, New York.
Thiede, T., Chievitz, E., and Christensen, B. C. (1964). Chlornaphazine as a bladder carcinogen. *Acta Medica Scandinavica*, **175**, 721–5.
Tucker, M. A., *et al.* (1987). Bone sarcomas linked to radiotherapy and chemotherapy in children. *New England Journal of Medicine*, **317**, 588–93.
Tucker, M. A., Coleman, C. N., Cox, R. S., Varghese, A., and Rosenberg, S. A. (1988). Risk of second cancers after treatment for Hodgkin's disease. *New England Journal of Medicine*, **318**, 76–81.
Valagussa, P., Santoro, A., Fossati Bellani, F., Franchi, F., Banfi, A., and Bonadonna, G. (1982). Absence of treatment-induced second neoplasms after ABVD in Hodgkin's disease. *Blood*, **59**, 488–94.
van Leeuwen, F. E., Somers, R., and Hart, A. A. M. (1987). Splenectomy in Hodgkin's disease and second leukaemias. *Lancet*, **ii**, 210–11.
Yunis, J. J. (1983). The chromosomal basis of human neoplasia. *Science*, **221**, 227–36.

Index

A-bomb survivors
  leukaemia risk in 14, 33
  malignancy in, after *in utero* irradiation 105
  and risk of stomach cancer 24
ABVD 60, 65, 164
acrolein
  metabolite of cyclophosphamide 1, 66, 85, 166
actinomycin 6, 50, 64, 165
actuarial risk
  as measure of cancer risk since treatment 4
acute leukaemia
  after alkylating agent therapy 65
  after chemotherapy for Behçet's syndrome 159
  after chemotherapy for myeloma 6
  after childhood cancer 112
  after Hodgkin's disease 26
  after ovarian cancer 26
  in trisomy 21 103
acute lymphocytic leukaemia
  in childhood, and subsequent malignancy 112
  in childhood, chemotherapy for 50
  after chloramphenicol use in childhood 110
acute myeloid leukaemia
  after alkylating agent therapy for benign disease 83, 85
  after chemotherapy 55
  after chemotherapy for myeloma 56, 157
  after Thorotrast exposure 38
acute non-lymphocytic leukaemia
  after alkylating agent chemotherapy 55, 64, 157
  after chemotherapy for Hodgkin's disease 116
  after chloramphenicol use in childhood 110
  after radiotherapy for cervical cancer 20
acyclovir 162
adducts: *see* carcinogen–DNA adducts
adriamycin 50, 60, 65, 164
alkylating agents
  as adjuvant therapy 7, 158
  for benign disease, and leukaemia risk 83
  carcinogenic potency of 2, 58, 164
  in childhood, and bone sarcoma risk 64

alkylating agents (*cont.*)
  for childhood leukaemia, and subsequent leukaemia risk 53
  and chromosomal damage 158
  dose-response for leukaemogenesis 26
  genetic susceptibility to carcinogenic effects of 117
  mechanism of action of 50, 167
  for ovarian cancer, and leukaemia risk 66
  versus radiation, as carcinogens 165
  versus radiation, as leukaemogens 61
  for retinoblastoma, and bone sarcoma risk 115
  for rheumatoid arthritis 1
  for systemic lupus erythematosus 1
alkylator score
  as index of total dose 58
  and leukaemia dose-response 115
alpha emitters
  ^{224}radium 36
  Thorotrast 38
AML: *see* acute myeloid leukaemia
analgesic mixtures
  and renal cancer 161
angiosarcoma
  after radiotherapy for breast cancer 6
ankylosing spondylitis
  and-leukaemia 14
  and stomach cancer 24
  treated by radiotherapy 30, 36, 45, 162
ANLL: *see* acute non-lymphocytic leukaemia
antidotes
  to carcinogenic effect of alkylating agents 166
antilymphocytic serum 162
antimetabolites 71, 167
antimitotics 167
appendectomy
  and Hodgkin's disease 139
appendix
  immunological function of 138
arsenic
  and skin cancer 161
arthroplasty
  of hip, and cancer risk 145
astrocytoma
  after cranial irradiation for leukaemia 6
  treated by radiotherapy 7
ataxia telangiectasia
  and cancer risk 79, 159

175

Index

augmentation cystoplasty: see cystoplasty
augmentation mammoplasty: see mammoplasty
autopsy series
  of colon cancer after cholecystectomy 133
  selection bias in 133
  of stomach cancer after gastrectomy 131
azathioprine
  versus alkylating agents, as carcinogen 83
  and breast cancer risk 90
  and non-Hodgkin's lymphoma 82
  for non-malignant disease, and cancer risk 79
  after transplantation, and cancer risk 71, 162

basal cell carcinoma
  after radiotherapy for medulloblastoma 119
BCNU 168
Beckwith–Wiedemann syndrome
  and adrenal tumours 104
  and hepatoblastoma 104
  and Wilms' tumour 104
Behçet's syndrome
  and genetic damage after chlorambucil 158
benign gynaecological disorders
  treated by radiotherapy 45, 162
bias
  ascertainment 128, 133, 143
  detection, in thyroid cancer surveillance 40
  recall, in case–control studies of childhood cancer 109
  selection 133
  telephone interviews 140
  under-reporting of exposure 141
bile acids
  metabolism after cholecystectomy 133
  reflux into stomach 131
biological markers
  of carcinogen exposure 158
  of genetic damage 160
bladder cancer
  after augmentation cystoplasty 147
  after azathioprine for ulcerative colitis 86
  after chemotherapy for polycythaemia vera 61
  after cyclophosphamide 1, 62, 64, 86, 166
bleomycin 60, 164
bone cancer: see osteosarcoma
bone marrow
  autotransplantation of 167
  toxicity of nitrosoureas to 168

brachytherapy: see radiotherapy
brain tumours
  after ^{131}iodine therapy 44
  after maternal use of diuretics 109
  after maternal use of phenobarbitone 108
  after radiotherapy for a previous brain tumour 7
  after radiotherapy for tinea capitis 110
  and risk of multiple tumours 120
breast cancer
  after augmentation mammoplasty 142
  after azathioprine for benign disease 90
  after cardiac pacemaker insertion 146
  after cholecystectomy 136
  contralateral, after radiotherapy for breast cancer 6
  contralateral, after radiotherapy to breast 42
  deficit of, after radiotherapy for metropathia 34
  after irradiation for thymic enlargement 43
  after multiple fluoroscopies 42
  protective effect of ovarian irradiation 20
  radiation-induced 1
  after radiotherapy for cervix cancer 19, 25
  risk dependence on age at irradiation 43
bullous pemphigoid 87

cancer
  after chemotherapy for childhood cancer 64
  radiation-induced 29
  after radiotherapy for ankylosing spondylitis 30
  after radiotherapy for metropathia haemorrhagica 34
cancer registries
  in case–control studies of cancer risk 140
  in cohort studies 14, 26, 78, 142, 145
  in surveillance of second cancer risk 154–6
  for retinoblastoma 118
  for vaginal adenocarcinoma 107
carcinogen-DNA adducts
  with cisplatinum 168
  for monitoring genetic damage 159
carcinogenesis
  mechanisms of 73
carcinogenic potency
  of chemotherapy agents 58, 165
  relative, of therapeutic strategies 153
cardiac pacemakers
  and cancer risk 146

case–control studies
  of cancer after appendectomy 138
  of cancer in childhood cancer
    survivors 64
  of cancer in twins 106
  of childhood leukaemia and maternal
    exposures 108
  versus cohort studies, for second cancer
    risk 25
  of colon cancer after cholecystectomy 133
  of immunization and childhood cancer 111
  of leukaemia after chemotherapy 55
  of leukaemia after Hodgkin's disease 60,
    165
  of leukaemia in childhood 110
  of leukaemia in childhood cancer
    survivors 115
  of neuroblastoma and maternal drug
    use 108
  of osteosarcoma after childhood cancer
    113
  of second cancers after cervix cancer 15,
    156
  of stomach cancer after gastrectomy 131
  for study of second cancer risk 155
  use of odds ratio in 4
case series
  versus population studies 113, 118
  for study of second cancer risk 154, 156
  use of percentages 3
CCRG: see Childhood Cancer Research
  Group
cell-killing effect
  after irradiation of pelvic marrow 35
  lack of, for thyroid cancer after
    irradiation 41
  of radiation, effect on leukaemia dose-
    response 14, 24
cerebral angiography
  use of Thorotrast in 38
cerebral irradiation
  for acute lymphoblastic leukaemia 6
  and astrocytoma 6
  and intellectual disturbance 6
cervix cancer
  in immunosuppressed transplant
    patients 74
  after immunosuppressive therapy for
    benign disease 88
  intracavitary radium for 13
  radiotherapy for 13
chemotherapy
  as adjuvant treatment 65, 163
  adverse effects of 2
  as carcinogen 153
  for childhood cancer, and subsequent
    malignancy 113
  and chromosomal aberrations 158

chemotherapy (*cont.*)
  for Hodgkin's disease 1, 26, 63
  interaction with radiotherapy 61, 165
  for cancer in transplant patients 78
  mechanisms of drug cytotoxicity 50
  for ovarian cancer 26
  therapeutic index of 166
Childhood Cancer Research Group 112,
  116, 120
childhood malignancy
  international variations in incidence
    of 101
chlorambucil
  for benign disease 71
  for ovarian cancer, and acute non-
    lymphocytic leukaemia 157
  relative carcinogenic potency of 165
chloramphenicol
  and childhood leukaemia 110
chlornaphazine 62, 164
cholangiocellular carcinoma
  after Thorotrast exposure 38
cholecystectomy
  and breast cancer 136
  and colon cancer 133
  indications for 133
choriocarcinoma
  treated with chemotherapy 50, 58
chromium
  in metal alloys 144
chronic fibrosis
  in foreign-body carcinogenesis 142
chronic lymphocytic leukaemia
  after radiotherapy for cervix cancer 21
  after Thorotrast exposure 39
cirrhosis
  after Thorotrast exposure 39
cisplatinum
  adverse effects of 166
  as animal carcinogen 65
  DNA adducts of, and cancer risk 160, 168
  with mesna, to reduce urinary tract
    toxicity 166
  for testicular teratoma 65
clinical trials
  in Hodgkin's disease, of less
    leukaemogenic therapy 65
  for study of second cancer risk, 8, 52,
    154, 157
CLL: *see* chronic lymphocytic leukaemia
cobalt
  in metal alloys 144
cohort studies
  of cancer after appendectomy 139
  of cancer after arthroplasty 145
  of cancer after Hodgkin's disease 26
  of cancer after irradiation for
    metropathia 34

cohort studies (*cont.*)
  of cancer after ovarian cancer 26
  of cancer after radiotherapy for cervix cancer 15
  of cancer after radiotherapy for tinea capitis 109
  versus case–control studies, for second cancer risk 25
  of colon cancer after cholecystectomy 134
  of osteosarcoma after childhood cancer 113
  of osteosarcoma after retinoblastoma 115
  of stomach cancer after gastrectomy 128
  for study of second cancer risk 155
colectomy
  prophylactic, in familial polyposis 127
  prophylactic, in ulcerative colitis 127
colon cancer
  adenocarcinoma, after ureterosigmoidostomy 147
  after cholocystectomy 133
  in familial polyposis 127
  and gallstones 136
  in ulcerative colitis 127
corrosion
  biological measures of 146
  of metal prostheses 144
cryptorchidism
  and prevention of testicular cancer 127
cumulative incidence rate 4, 118
cumulative risk
  of cancer after radiotherapy for retinoblastoma 118
  of cancer after Wilms' tumour 117
  in childhood cancer survivors 112
  of osteosarcoma after childhood cancer 115
cyclophosphamide
  and bladder cancer 1, 64, 85, 166
  and leukaemia 157
  for non-malignant disease, and cancer risk 79
  synergism with irradiation in carcinogenesis 115
cyclosporin
  and cancer risk 75
cystoplasty
  and bladder cancer 147
cytokines 75, 79
cytomegalovirus
  in Kaposi's sarcoma 74
cytotoxic agents
  side-effects of 71
cytotoxicity
  mechanisms of 50

dacarbazine 60, 65, 164, 168
deoxycholic acid: *see* bile acids
dermatomyositis
  and Kaposi's sarcoma 87
  versus paraneoplastic syndromes 81
diethylstilboestrol (DES)
  as transplacental carcinogen 107
  and vaginal adenocarcinoma 107, 161
diphenylhydantoin
  as transplacental carcinogen 108
disease-free survival
  in risk–benefit evaluation 2, 8
diuretics
  and brain tumours in offspring 109
dose rate
  in radiation carcinogenesis 27
dose-response
  for bone sarcoma after radium-224 therapy 37
  for brain tumours after scalp irradiation 110
  for breast cancer after irradiation of thymus 43
  for cancer after chemotherapy 163
  for childhood cancer after *in utero* irradiation 105
  for childhood leukaemia after chloramphenicol 110
  for genetic damage after alkylating agent therapy 159
  for leukaemia after alkylating agents 26, 157
  for leukaemia after radiotherapy 14, 155
  for leukaemia by alkylator score 115
  models, for radiation leukaemogenesis 22
  for osteosarcoma after chemotherapy 113
  for osteosarcoma after radiotherapy 113
  relationships for second cancer risk 158
  for skin cancer after PUVA 160
  for stomach cancer after radiotherapy 24
  for thyroid cancer after irradiation of thymus 41, 109
  for thyroid cancer after radiotherapy in childhood 116
doxorubicin
  and luekaemia risk 115
drug structure
  modification of 167
drug targeting
  with monoclonal antibodies to tumour receptors 167

EBV: *see* Epstein–Barr virus
ELISA 159
embryonal tumours 101
endometrial cancer
  after cervix cancer 20

Index

endometrial cancer (*cont.*)
    and post-menopausal use of oestrogens 161
    after tamoxifen for breast cancer 7
epilation
    radiation-induced, for tinea capitis 41
Epstein–Barr virus
    and lymphoid malignancy in transplant patients 73, 75
    mechanism of carcinogenesis in immunosuppression 75
N-ethylnitrosourea 107
European Organization for Research and Treatment of Cancer 156, 161
Ewing's sarcoma
    and osteosarcoma 115

familial polyposis
    prophylactic colectomy in 127
Fanconi's anaemia
    and cancer risk 159
fetal hydantoin syndrome 108
fibroadenomatosis
    treated by radiotherapy 42
fibrosarcoma 37
fluoroscopy
    in imaging for pneumotherapy 42
fluorouracil 164
foreign body
    carcinogenesis from 141
    and sarcoma 145
fractionation
    of breast irradiation, effect on breast cancer risk 43
    of radiotherapy, to reduce cancer risk 163
    of radium-224 therapy, effect on sarcoma risk 37

gallstones
    and colon cancer 136
ganglioneuroblastoma
    after maternal diphenylhydantoin 108
gastrectomy
    and other cancer 131
    and stomach cancer 128
genetic susceptibility
    in children, to malignancy after treatment 117
germ cell mutations
    after treatment of childhood malignancy 117
Gorlin's syndrome
    and genetic susceptibility to carcinogenesis 117
    radiosensitivity in 118

haemangiosarcoma
    after Thorotrast exposure 38
haemodialysis
    to reduce toxicity of chemotherapy 167
haemopoietic neoplasms
    after hip arthroplasty 145
haemorrhagic cystitis
    after cyclophosphamide 62, 66, 85, 166
haloethylnitrosoureas 168
hepatoblastoma
    in Beckwith–Wiedemann syndrome 104
hepatocellular carcinoma
    after immunosuppressive therapy for rheumatoid disease 89
Herpes zoster
    treated with radiotherapy 1
highly selective vagotomy
    and stomach cancer 131
hip replacement: *see* arthroplasty
Hodgkin's disease
    after appendectomy 139
    in childhood, and cancer risk 101, 112
    disease-free survival in 1, 5
    and risk of second cancer 156
    in systemic lupus erythematosus 81
hyperthyroidism
    treated with iodine-131 44
hypochlorhydria
    after gastrectomy 131

IARC: *see* International Agency for Research on Cancer
ileal conduits
    for urinary diversion, and cancer risk 147
immunization
    as protective factor against cancer 110
immunodeficiency
    acquired, and cancer risk 79
    congenital, and cancer risk 74
    and lymphoma risk 137
immunosuppressive therapy
    for autoimmune disease 71
    as carcinogen 153
    and cervix cancer 88
    and lymphoma 76, 162
    for organ transplantation, and cancer risk 71, 73
    and viruses in post-transplant malignancy 162
immunosurveillance theory 73, 138
incidence rate
    average annual 3
    cumulative 4
    versus proportion, as measure of cancer risk 147
intellectual disturbance
    after cerebral irradiation for cancer 6

International Agency for Research on
  Cancer 75, 101, 155, 161
Inter-Regional Epidemiological Study of
  Childhood Cancer 111
iodine-131
  for thyrotoxicosis 44
ionizing radiation
  prenatal exposure to, and childhood
    cancer 104
IRESCC: see Inter-Regional
  Epidemiological Study of Childhood
  Cancer
ischaemic heart disease
  after splenectomy 137

Kaposi's sarcoma
  after immunosuppressive therapy for
    benign disease 86
  after transplantation 78
keratoacanthoma 88
Knudson
  mutation model for retinoblastoma 102

Late Effects Study Group 112, 120, 156
latent period
  for bladder cancer after
    cyclophosphamide 62
  for cancer after transplantation 78
  of carcinogenesis after chemotherapy 64
  in chemotherapy-induced leukaemia 56
  in radiation-leukaemia models 24
  in Thorotrast-induced cancer 38
  for tumours induced by
    immunosuppression 74
leiomyosarcoma
  of uterus after pelvic irradiation 36
LESG: see Late Effects Study Group
leukaemia
  in A-bomb survivors 14
  in ankylosing spondylitis 14
  after chemotherapy 52
  after chemotherapy for breast cancer 5,
    7, 60
  after chemotherapy for Hodgkin's disease
    26, 53, 58
  after chemotherapy for leukaemia in
    childhood 53
  after chemotherapy for non-Hodgkin's
    lymphoma 61
  after chemotherapy for ovarian
    cancer 26, 60, 66
  after childhood cancer 115
  after Ewing's sarcoma 115
  after Hodgkin's disease 5, 115
  after Hodgkin's disease in childhood 112
  after *in utero* irradiation 104

leukaemia (*cont.*)
  after iodine-131 for thyrotoxicosis 44
  after maternal use of narcotics 109
  radiation-induced 1, 14, 29, 155
  after radiotherapy for ankylosing
    spondylitis 30, 33
  after radiotherapy for breast cancer 6
  after radiotherapy for cervix cancer 21
  after radiotherapy for Hodgkin's
    disease 53
  after radiotherapy for metropathia
    haemorrhagica 34
  after radiotherapy in children 109
  after radium-224 therapy 36
  after splenectomy 137, 161
  after Wilms' tumour 115
leukaemogenesis
  interaction between radiotherapy and
    chemotherapy 61
leukaemogenic potency
  comparison of cytotoxic agents 57
Li–Fraumeni syndrome
  and familial risk of cancer 120
  and genetic susceptibility to
    carcinogenesis 117
life-table probability
  use in risk estimation 4
lithocholic acid: see bile acids
liver cancer
  after Thorotrast exposure 38
lung cancer
  after gastrectomy 131
  after radiotherapy for cervix cancer 15
  after radiotherapy for Hodgkin's
    disease 63
lymphangiosarcoma
  after mastectomy 161
lymphoedema 161

mammography
  effect of silicone implants on efficacy
    of 143
  low-dose, and breast cancer risk 25
mammoplasty
  augmentation, cancer risk after 142
  reduction, cancer risk after 143
mastectomy
  radical, and lymphangiosarcoma 161
mastitis
  postpartum, treated by radiotherapy 1, 42
Medical Research Council 157
medullary thyroid carcinoma
  in multiple endocrine neoplasia 127
medulloblastoma 119
melphalan
  and acute myeloid leukaemia 157
  and chromosomal rearrangements 159

meningioma 7
menopause
 radiation-induced 20, 33, 34
mesenchymoma 108
mesna (mercaptoethane sulphonate)
 use with cyclophosphamide 1, 66, 166
mesothelioma
 after Thorotrast exposure 39
metastases
 misclassification as second cancer 19
methotrexate 58, 71, 89, 164
methoxsalen 159, 160
methyl–CCNU: see semustine
methylphenobarbitone 108
metropathia haemorrhagica
 treated by radiotherapy 34
microglioma 76
molybdenum
 in metal alloys 144
monoclonal antibodies
 to DNA-adducts, use in biomonitoring 159
 use in drug targeting 167
MOPP 58, 60, 65, 164
multiple endocrine neoplasia
 prophylactic thyroidectomy in 127
multiple myeloma: see myeloma
mustine 71
MVPP 60
myasthenia gravis
 treated by thymectomy 138
myelodysplastic syndrome
 after chemotherapy 55
 after chemotherapy for myeloma 56
 after splenectomy 137
myeloma
 and acute leukaemia after chemotherapy 6
 after radiotherapy for cervix cancer 20

naphthylamine 62
National Cancer Institute 8, 156
neuroblastoma
 after maternal diphenylhydantoin 108
nitrogen mustard 58, 164
N-nitroso carcinogens
 and cholecystectomy, in gastric cancer risk 133
 from gastric remnant 133
nitrosoureas
 and leukaemogenesis 158
 toxicity to bone marrow of 168
 as transplacental carcinogens 107
non-Hodgkin's lymphoma
 cerebral, after transplantation 76

non-Hodgkin's lymphoma (cont.)
 after cytotoxic therapy for rheumatic disease 82
 after renal transplantation 72
 in Sjögren's syndrome 81

odds ratio
 as estimate of relative risk 4
oestrogens
 and endometrial cancer 161
oncogenes 103, 159, 169
orchidectomy
 in cryptorchidism 127
orchidopexy
 in cryptorchidism 127
orthopaedic prostheses: see arthroplasty
OSCC: see Oxford Survey of Childhood Cancer
osteosarcoma
 after chemotherapy for Hodgkin's disease 116
 after Ewing's sarcoma 115
 after genetic retinoblastoma 118
 after radiotherapy for breast cancer 6
 after radiotherapy for retinoblastoma 27
 after radiotherapy in childhood 113, 156
 after radium-224 therapy 36
 after Thorotrast exposure 39
 after treatment of retinoblastoma 112, 115, 118
ototoxicity 166
ovarian cancer
 after cervix cancer 20
 and leukaemia after chemotherapy 53
 and leukaemia after melphalan 6
oxazaphosphorine 166
Oxford Survey of Childhood Cancer 104–6, 108, 111

parotid tumours
 after radiotherapy for tinea capitis 110
partial gastrectomy: see gastrectomy
pemphigus vulgaris 87
penicillamine 80
person-years
 of observation, use in risk estimation 3
phagocytes 74
phenacetin
 and renal cancer 161
phenobarbitone
 and brain tumours 108
 as transplacental carcinogen 108
phenylbutazone 80

phenytoin: *see* diphenylhydantoin
photoadducts
   with DNA, after PUVA therapy 160
pneumonia
   after splenectomy 137
pneumotherapy 42
polyarteritis nodosa 81
polymyositis 87
postpartum mastitis: *see* mastitis
prednisolone 71, 75, 87, 88
preleukaemia 55, 158
prevention
   primary, with surgery
   of therapy-related tumours 169
procarbazine 58, 65, 164
prostate cancer
   after vasectomy 140
PUVA
   and DNA photoadducts 160

radiation
   adverse effects of 29
radiation dose
   to breast from fluoroscopy 42
   to breast from irradiation of thymus 43
   cell-killing effect 14
   for cervix cancer, non-uniform distribution of 27
   to individual organs 13
   to liver from Thorotrast exposure 38
   low-dose, effects on cancer risk 29
   phantom simulation of 14
   in screening, and study of cancer risk 29
   to skeleton from radium-224 36
   to thyroid gland from irradiation of thymus 40
radiosensitivity
   of female breast 42
radiotherapy
   for ankylosing spondylitis 45
   for benign breast disease 42
   for benign conditions of the head and neck 40, 45
   for benign gynaecological disorders 33, 45
   for benign tumours 7
   as carcinogen 153
   for cervix cancer 13
   for childhood cancer 112
   for enlargement of the thymus 43
   for Herpes zoster 1
   for Hodgkin's disease 26
   interaction with chemotherapy 61, 165
   for osteoarthritis 1
   for ovarian cancer 26
   for postpartum mastitis 1, 42
   for tinea capitis 40, 41

radiotherapy (*cont.*)
   for tuberculosis 36
   for whooping cough 40
radium-224
   intravenous injection of 36
reduction mammoplasty: *see* mammoplasty
relative rate
   as method of comparing risks 4
relative risk
   of cancer, estimated from clinical trials 157
   as method of comparing risks 4
renal cancer
   after radium-224 therapy 36
repair enzymes
   and carcinogenesis 65
retinoblastoma
   hereditary versus sporadic 102
   Knudson model for 102
rheumatoid arthritis
   predisposition to lymphoma in 80
   treated with alkylating agents 1, 5
   treated with immunosuppressive therapy 71
ringworm: *see* tinea capitis
risk coefficient
   for radiation carcinogenesis in thyroid gland 41, 109
risk–benefit evaluation
   after chemotherapy 164
   disease-free survival in 2, 8
   in surgery 127
   and quality of life assessment 2, 5
   quantitative approach in cancer therapy 3

sarcoma
   after radiotherapy for benign gynaecological disorders 36
   after radiotherapy for breast cancer 6
   soft tissue, in childhood, and cancer risk 112, 120
   after subcutaneous silicone implant 142
SCE: *see* sister chromatid exchange
screening, effects of radiation exposure from 29
second cancer risk
   treatment versus genetic predisposition 113
SEER: *see* Surveillance, Epidemiology, and End Results programme
semustine
   bone-marrow toxicity of 168
   and leukaemia 58, 157, 158
silicone-gel prostheses
   and cancer risk 142
Sippel's syndrome: *see* multiple endocrine neoplasia

sister chromatid exchange
  after chlorambucil for Behçet's syndrome 158
  as marker for genetic damage 159
Sjögren's syndrome
  and non-Hodgkin's lymphoma 81
skin cancer
  squamous, after transplantation 72, 74
SLE: *see* systemic lupus erythematosus
smoking
  cessation of, in Hodgkin's disease 167
soft tissue sarcoma
  and multiple tumours 120
  after retinoblastoma 112
solid tumours
  after *in utero* irradiation 104
spleen
  immunological function of 137
splenectomy
  and cancer risk 137
  in Hodgkin's disease 137
  and ischaemic heart disease 137
  and leukaemia 137, 161
  and myelodysplastic syndrome 137
  and pneumonia 137
stomach cancer
  after cervix cancer 24
  after gastrectomy 128
surgical procedures
  anaesthetic complications of 127
  and cancer risk 161
surveillance
  of cancer risk using biological markers 160
  of cancer survivors 153, 154
  of children of cancer survivors 117
  of dysplastic naevi in Hodgkin's disease 167
  after gastrectomy 128, 131
  methods of case ascertainment in 158
  after ureterosigmoidostomy 148
Surveillance, Epidemiology, and End Results programme 156
syntomycin: *see* chloramphenicol
systemic lupus erythematosus
  and bladder cancer after cyclophosphamide 86
  treated with alkylating agents 1

tamoxifen
  for prevention of breast cancer 7
  preventive efficacy of 8
  for treatment of breast cancer 7
teletherapy: *see* radiotherapy
temporal patterns
  of bladder cancer after cyclophosphamide 62

temporal patterns (*cont.*)
  of leukaemia after chemotherapy 56, 156
  of skin cancer after transplantation 76
testicular cancer
  less intensive treatment for 167
  primary prevention of, in cryptorchidism 127
  treated with cisplatinum 65
  after vasectomy 141
testosterone
  and prostate cancer 140
therapeutic index
  of anti-cancer drugs 166
thorium dioxide: *see* Thorotrast
Thorotrast
  body distribution of 38
  in contrast radiography 37
thymectomy
  for myasthenia gravis 138
  for thymoma 138
thymus
  radiotherapy for enlargement of 40, 43
thyroid adenoma
  after radiotherapy to thymus 40
thyroid cancer
  effect of age at irradiation on risk of 116
  after iodine-131 for thyrotoxicosis 44
  after radiotherapy for neuroblastoma 116
  after radiotherapy for tinea capitis 41, 110
  after radiotherapy in childhood 116
  after radiotherapy to head and neck 40
  after radiotherapy to thymus 109
thyrotoxicosis
  treated with radiation 45
tinea capitis
  and cancer risk after radiotherapy 109
  treated by radiotherapy 162
tonsillectomy
  and cancer risk 138
tonsils
  immunological function of 138
transplacental carcinogenesis
  in animals 106
  by diethylstilboestrol 107
  by diphenylhydantoin 108
  in humans 107
  by nitrosoureas 106
  by phenobarbitone 108
transplantation
  autologous, of bone marrow 167
  and cancer risk 72
  effect of previous cancer on cancer risk after 78
  immunosuppression as carcinogenic mechanism in 73, 162
  and Kaposi's sarcoma 78
  role of EBV in carcinogenesis after 75
  use of immunosuppressive therapy in 71

treatment regimes
  modification of, for cancer risk reduction 153, 161
  recording by cancer registries 155
treosulphan 58, 164
triazenes 168
trisomy 21
  and acute leukaemia 103
tuberculosis
  treated by pneumotherapy 42
  treated by radiotherapy 36
tumour suppressor genes: *see* oncogenes
twins
  cancer risk in, after *in utero* irradiation 106

ulcerative colitis
  associated with ankylosing spondylitis 30
  prophylactic colectomy in 127
ultrasound
  versus radiology, in obstetrics 121
UNSCEAR 56, 105
ureterosigmoidostomy
  and colon cancer 147
  for urinary diversion 147

vaccination
  viral, against post-transplant cancer risk 79
vaginal adenocarcinoma
  after maternal diethylstilboestrol 107

vasectomy
  and prostate cancer 140
  and testicular cancer 140
vinblastine 60, 164
vincristine 58, 60, 164
viruses
  and carcinogenesis after transplantation 73
  and immunosuppression in post-transplant malignancy 162
Vitallium 145
von Recklinghausen's syndrome
  and genetic susceptibility to carcinogenesis 117
  multiple tumours in 112, 120

Waldenström's macroglobulinaemia 81
Wegener's granulomatosis
  and bladder cancer after cyclophosphamide 86
whooping cough
  treated by radiotherapy 40
Wilms' tumour
  11p deletion syndrome 104
  and Beckwith–Wiedemann syndrome 104
  incidence of 101
  and subsequent malignancy 112, 115, 117
  treated with chemotherapy 50
Wiskott–Aldrich syndrome
  and cancer risk 79

zidovudine 79

THE LIBRARY
UNIVERISTY OF CALIFORNIA, SAN FRANCISCO
(415) 476-2335

THIS BOOK IS DUE ON THE LAST DATE STAMPED BELOW

Books not returned on time are subject to fines according to the Library Lending Code. A renewal may be made on certain materials. For details consult Lending Code.

14 DAY

NOV 26 1993